Leading Project-Driven Change

Richard C. Bernheim, PMP, MBA

First Edition

Multi-Media

Oshawa, Ontario

Leading Project-Driven Change
by Richard C. Bernheim, PMP, MBA

Managing Editor:	Kevin Aguanno
Copy Editor:	Susan Andres
Typesetting:	Peggy LeTrent
Cover Design:	Troy O'Brien
eBook Conversion:	Charles Sin and Agustina Baid

Published by:
Multi-Media Publications Inc.
Box 58043, Rosslynn RPO
Oshawa, ON, Canada, L1J 8L6

http://www.mmpubs.com/

Paperback	ISBN-13: 978-1-55489-115-3
Adobe PDF ebook	ISBN-13: 978-1-55489-116-0

Published in Canada. Printed simultaneously in Canada, the
United States of America, Australia, and the United Kingdom.

CIP data available from the publisher.

Table of Contents

I dedicate this book to the late Rita Mulcahy of RMC. I hold deep respect for her and am thankful for the help her organization has given me.

—*Richard C. Bernheim*

Preface

I felt compelled to write this book as a recent follow-up to my previous work, *People-Centric Project Management*. In part 1 of this book, I share five short fables with the reader. My point is neither to emphasize technology nor to portray any particular period, company or industry information, or any such details. Rather, these five short fables are meant to help the reader better appreciate the difficult job of project manager in the real world. In addition, these stories help set the stage for part 2 of this book, which addresses how best to achieve *Project-Driven Change Leadership*.

Each short fable is about a particular project situation. Although technology and process (methods and tools) elements play a role in each of these five tales, neither is a driver for how each project situation ultimately plays out. Instead, each short fable is presented to help you understand how the project outcome is achieved. These project results have far more to do with the people elements of project management than anything else. These five short fables are meant to describe real-life *Project-Driven Change Leadership* in action, the ability to handle resistance to change, dealing with various people challenges that arise in project leadership, and working to persevere despite many distractions and hardships endured. They are all fictional situations inspired by true stories.

Further, following the entire presentation of all five short fables, I analyze the Lessons Learned from each and overall. Then, in part 2 of this book, I give the reader detailed advice having to do with the elements involved with how best to achieve *Project-Driven Change Leadership*:

- Aligning your actions with your values
- The important roles of prioritizing and focusing
- Learning, challenging, and growing
- Setting and reviewing your goals
- The role of measuring your results
- Specific "best practice" ideas about *Project-Driven Change Leadership*

All these ideas are defined, explained, and summarized both by chapter and then for all of part 2.

Last, in part 3, titled "One Large Fable," I share another story with the reader to illustrate a "worst-case" project scenario through a series of mistakes and misunderstandings from the start of the endeavor. These project management errors are obvious and many. Some are funny, whereas others are not, and it all is meant to give the reader a solid learning experience through storytelling.

The book is summarized, ending positively. The bottom-line message is that although leading project-driven change is usually challenging, it is doable.

Acknowledgments

First, this book would not be possible without the strong support of my wife and family. Further, without the many information-technology consulting opportunities I have experienced over the last thirty years, I would lack both the knowledge and experience necessary to create this book. Finally, without my involvement in and passion for the profession of project management, I would lack any interest in sharing my experiences with others.

Although lecturing to people has value, I believe storytelling is even more effective in driving home what matters most. I believe strongly that the project management community is too focused and enamored of improved methods, tools, and technology at the expense of something far more significant in this profession. Unless project managers know how to deal well with people, they jeopardize the effort with which they are involved. The researchers studying project failure bear out this truth repeatedly.

Last, by practicing project management, I have become exposed to an up-and-coming sister profession known as change management. I now believe you need to combine the ideas and best practices of both to be successful with information technology endeavors. Project and change management are difficult to achieve because they cause people to think and do things very differently from how they have done them before. To be effective, one must understand both professions and apply the lessons learned by so many who have toiled beforehand to today's challenging projects and those in the future.

Richard C. Bernheim, PMP, MBA
October 2010

Part One:

Five Short Fables

Leading Project-Driven Change

CHAPTER ONE

A Bit of a Technological Change

Background Information

A company manufactured transmitters and receivers for both domestic and international suppliers of military aircraft and naval vessels. This company had a small division that produced printed circuit boards (PCBs) used in these products and many others. Half these PCBs were sold to external customers, while the other half were provided to its sister divisions on a cost-plus or transfer-pricing basis. This division had been in business for a long period, and more recently, it had been consistently operated at a solid 10% pretax profit.

One of its key work centers, the drilling operation, was a significant bottleneck in throughput for the entire manufacturing process. One person operating a manual drill press drilled all the holes required for each PCB produced, according to each customer's specifications. This operation was very tedious and too frequently prone to human error because the operator reviewed each hole drilled. Further, changing drill bit sizes required extra effort to drill all different sizes of holes, which also introduced potential mistakes. The physical space in which this work center had to operate was also too small to handle all the PCBs coming in to get their holes drilled from the previous work center, the cutting operation. So, as a result, this division's parent company decided to investigate a new cutting-edge piece of equipment to increase this work center's throughput, and thus its output, and to improve its quality. In time, a cost-benefit analysis was performed, and not long thereafter, a proper piece of equipment was discovered and bought a short while later.

To accommodate this large machine, the current room used by the drilling operation had to be enlarged before this new piece of equipment could be installed, resulting in an additional cost. Once this was done, and its manufacturer deemed the machine operational, it could drill holes in as many as sixteen PCBs at once, while automatically changing drill bits to produce all the required and different hole sizes. For this to happen, an electronically programmed tape with the customer's detailed specifications first had to be created and then provided to the machine.

Now, instead of PCBs stacking to have their holes drilled, they accumulated afterward, awaiting the next operation in manufacturing. The drilling work center still required one operator, but this person's work was now different and more productive, thanks to this new piece of expensive equipment. All it required was electricity while it was used. Product quality was now much better as well. Everyone involved was pleased with the early results achieved.

Then, a problem arose that no one expected. Suddenly, the consistent month-end profitability results became very erratic. One period, they would rise well over the 10% pretax norm only to then fall well below this threshold the next period for no apparent reason. So, instead of the customary annual physical plant inventory, corporate executive management now started monthly physical inventories. Just as with profits, one month, work-in-process inventory valuation would jump, to be followed the next month by a sharp drop. The lack of consistency without good explanation resulted in two responses from this division's parent company.

First, the plant controller, Jim, was ordered to investigate this situation and provide a full and detailed report within sixty days. After doing so, Jim found no particular reason for this behavior, except that it persisted. Customer orders and sales remained stable; PCB physical inventory counts were consistent; and manufacturing costs were reasonable compared with sales. So, as a result, Jim was shortly replaced from his position. How this was handled involved swapping the assistant controller, Ted, who was at a nearby and larger sister division, with Jim. Jim was demoted, whereas Ted was promoted with his first key assignment being to find this crucial financial matter's underlying cause and correct it immediately. Concurrently, the PCB division plant manager, Bob, was told to do everything possible to help Ted with this important assignment. The parent company president and corporate controller informed both Bob and Ted that unless things returned to the previous stable financial position, they would consider selling or even closing this PCB division. PCBs could be obtained elsewhere without problem, if need be, from many other good domestic suppliers.

So, Bob and Ted went to work on this critical problem right away. They agreed that Ted needed to learn the entire business operation in detail to be most effective, not just in this first assignment, but for the long term, assuming things went well. Bob gave Ted a thorough plant tour, introducing him to all division employees. Once this was completed, Ted began to meet with each work-center supervisor to delve much deeper into each operation in the PCB manufacturing process.

This self-initiated effort took Ted several weeks. Ted took very detailed personal notes, which when finished, were passed along to Bob for his review and approval. Other than a few minor wording clarifications, Bob told Ted that his notes were accurate and complete.

Now, once this was finished, Ted turned his focus on the small financial staff and to the division's accounting records and financial statements. He worked on these two things long and hard to become very familiar with them all. The sudden transition from Jim to Ted was somewhat disconcerting to the financial staff. Ted moved swiftly to address their questions and concerns because Jim had held the controller position previously for more than 10 years. Ted also got into all current and recent accounting records with their corresponding financial statements in detail. If he had a question that neither the financial staff nor Bob could answer, he could query Jim about it. Jim was made available, including in person, as Ted required, and he did everything possible to help Ted.

So, within ninety days, Ted felt comfortable enough to announce that he was ready to begin to probe the erratic work-in-process inventory and profitability situation that continued during this period. He focused simultaneously on two key areas: the work-in-process inventory valuation process and the month-end financial statements, mainly the balance sheet and income statement. These financial statements were provided subsequently and periodically to the parent company for a corporate financial reporting consolidation.

First, Ted examined the month-end physical inventory counting process with the corresponding procedures in some detail. He discovered no irregularities of any kind in this process. He also challenged the financial staff and all accounting records, seeking any errors or missing data. Because most accounting records and the creation of periodic financial statements and corresponding reports with all relevant analysis were manual, many paper documents and mathematical calculations were involved.

Ted tracked and carefully monitored everything. Up to this time, nothing was lacking, missing, or in error. Ted made many suggestions for improving these various processes, but none of these altered the work-in-process inventory valuations and the month-end financial statement results, which just continued erratic during this period. Ted involved

Bob in his examination of the work-in-process inventory preparations, including the physical counting process, but Bob saw nothing unusual. Ted and Bob were somewhat disturbed that nothing jumped out at them after counting the entire work-in-process inventory carefully and thoroughly.

Next Steps

Once the physical inventory counting process was over, it next fell to Ted to use this information to value the work-in-process inventory for financial statement reporting. From his predecessor, Jim, who had passed along to Ted the entire manual valuation process including various worksheets, came the final periodic financial results. Those who preceded Jim as this division's controllers passed this process and the worksheets to him. Because the work environment had been stable for so many years, there was no need to challenge or to change this process, which Jim hadn't, and for now, Ted followed.

Work-in-process inventory valuation entailed the use of three manual worksheets on which the first physical inventory count data were recorded. Then, some basic mathematical calculations were performed, and interim results were transposed on to the next worksheet manually. The last worksheet produced the financial results used to create the month-end journal entries and, ultimately, the income statement and balance sheet.

Ted followed the detailed instructions he received from Jim carefully and methodically. He double-checked all mathematical computations in each worksheet. He double-checked for any transposition errors when posting his interim results from one worksheet to the next worksheet because it was manual. This entire effort took some three hours without interruptions. Then, shortly thereafter, the period-end financial statements became available for reporting and further analysis.

So even before Bob could see the division's final numbers, Ted could reflect on the results shown him. The income statement revealed a big drop in both work-in-process inventory and pretax profit from the prior month. Ted focused on these numbers and saw little change in sales and costs to justify such a steep decline. He finally visited Bob to share the financial statements and his first analysis of the results achieved.

Bob, like Ted, was completely befuddled. They discussed the situation for some time, but found no clear explanation. Ted was now responsible to pass along the division's financial statements to the corporate controller, George, for a corporate-wide financial reporting consolidation. But both Ted and Bob knew they would now be held accountable for the poor numbers reported when they were transferred by Ted to George over at corporate.

The corporate president and controller soon called on Ted and Bob to get more information and insight into this situation. Neither Ted nor Bob could offer them any good explanation. The corporate president reminded them of his displeasure with these poor financial results and that the parent company considered selling or closing this PCB division unless things got dramatically better and fast and then stayed that way consistently for a long while.

Ted and Bob got the message loud and clear. They understood how frustrated the parent company executives were about this financial situation, especially after they spent so much money on the new automated drilling machine, including the drilling operation room changes required because of its large size. They recommitted themselves to finding the cause of this important matter and to correcting it as soon as possible.

So now, every end of week, they carefully examined various key business indicators, such as incoming customer orders, sales, and costs, for any irregularities or significant fluctuations. They monitored the factory floor, looking for work-in-process inventory spikes and drops by reviewing things as best they could. Then, once month-end approached, they carefully prepared for the upcoming physical inventory of work-in-process. They double-checked the procedures beforehand and during the physical inventory counting.

Plant security and surveillance were enhanced because one raw material used in producing these PCBs was gold because it was such a good conductor of electricity. When PCBs were found somehow defective after they had the gold placed on them, they were reclaimed to extract and then sell this gold as a source of other income.

Scrap rates were carefully measured, but they never indicated a problem. If anything, given the new automated drilling machine's effectiveness, scrap rates now ran consistently well below the historical norms of this manufacturing process at this division. Ted carefully took the physical inventory counts and ran them through work-in-process inventory valuation using the three manual worksheets. He double-checked each step along the way, including all mathematical computations involved in this tedious process. After obtaining the results and then completing the financial statements, he discovered a dramatic increase in both the value of the work-in-process inventory and pretax profit.

Resolution

Now, although this was good news, especially following the previous period's poor financial results, neither Ted nor Bob expected this, given the stability in all other key business indicators during the month. Sales were flat; customer orders were slightly down; and scrap was stable and on the low side. In addition, costs were a bit higher compared with the prior month's results. Nothing seemed to explain this good fortune.

So, again, the next period, the financial results took another big nosedive. Everyone's patience by now was wearing thin. Ted didn't wait to be told again how serious this matter was. He sprang into action by working backward from the financial results reported in the financial statements to the work-in-process inventory valuation to the other accounting source documents. Ted traced every detail and double-checked each calculation. He focused intently on each piece of information, but for some unknown reason other than his intuition, he seemed more attracted to the three manual worksheets used to value the work-in-process inventory.

Ted somehow sensed something in this manual process was the reason the financial outcome was what it was. The reason he felt this way was that every other financial record was verifiable, but with this process, he had to rely completely on the various factors in the three manual worksheets he inherited from his predecessor controllers. It all apparently seemed both logical and reasonable, but it all relied on only historical use information.

Ted began to tear into each factor in these three manual worksheets. He went back to the worksheets he used in the previous three accounting periods and compared the results for each factor. Of the many factors included in these three manual worksheets, the one factor showing the most variability was the one in the second worksheet that placed a value on the number of drilled holes. This factor was a number with six places to the right of the decimal point, thus portraying it as a precise factor. Ted now went to visit Bob about his inkling.

Bob indicated to Ted that he was aware that drilled holes had a value factor associated with it. Ted then probed deeper into this by asking Bob about the number of holes used to multiply by this precise factor. He mentioned that, during each physical inventory, he did not recall anyone being asked to count the number of drilled holes in each PCB. Bob explained that the number of holes was an estimate created many years earlier from an industrial engineering study undertaken at this PCB division. Then, Ted asked Bob why each drilled hole had an identical value when hole sizes varied a good bit. Bob responded that the industrial engineer accounted for this aspect and concluded that, in valuing drilled holes, there was no discernible or material difference. Further, the time required for changing drill bit sizes was incorporated into this financial valuation factor. Ted considered all this information and then asked Bob to be excused so he could return to his office to reflect further on this new information.

Back in his office, Ted got back to his analysis. He realized that the factor used to value the number of drilled holes was created during the long period when the drilling operation was manual and thus before the new automated machine came online. He pulled out the work-in-process inventory valuation worksheets for the last eight months since this drilling machine had become operational. He recorded the number of drilled holes and the associated financial results after the financial valuation factor had been applied on a thirteen-column accounting spreadsheet. Then, he dug out the final total work-in-process inventory valuation and pretax profit or loss figures for each of the last eight month-end financial reporting periods. He also averaged these results to determine the extent of variation they exhibited.

Next, with the financial staff's help, he went back even further into the older accounting records to the same results dating six months before the new automated drilling machine became available to this division. Again, he recorded the number of drilled holes and their financial results following the application of the valuation factor, final work-in-process inventory valuation, and pretax profit or loss figures. He also averaged each of these numbers for these six accounting periods. Then, he compared all the results and discovered a correlation between an increase in the financial valuation of drilled holes and an increase in work-in-process inventory valuation and pretax profit. When the financial valuation of drilled holes decreased, so did work-in-process inventory valuation and pretax profit.

Ted now told these findings to Bob who had just gotten off the phone with the company president who was upset about the steep decline in pretax profit from the preceding month's financial results. Ted explained the data in his spreadsheet to Bob and the trend he had discovered. They then spent some time reviewing and analyzing this information. Bob now felt Ted was onto something important.

He realized that by using the new automated drilling machine, instead of the former manual drill press, that the factor used to value drilled holes, whereas seeming precise, was now suspect. The new machine's ability to drill holes in as many as sixteen PCBs simultaneously, with no downtime for drill bit changeovers when different hole sizes needed to be drilled, meant the hole valuation factor used was distorting the total work-in-process inventory valuation result and, thus, significantly affecting profit.

As they discussed this further, Ted asked Bob again to be excused. He returned to his office to try another idea he now had. What if drilled holes were no longer valued because the new factor to be used was likely to be so small as not even to matter? Ted went back over the work-in-process inventory valuation worksheets; only this time, he excluded valuing drilled holes for the most recent eight-month period to see what difference it would make in total work-in-process inventory valuation and pretax profit. He discovered that, by ignoring the valuation for holes drilled, total work-in-process inventory valuation each period remained stable, which correspondingly kept pretax profit very close to its long-term historical division norm of 10%.

Ted then raced back to Bob's office to reveal his findings. After explaining it all, Bob had to admit it made sense. Ted then asked Bob how this practice could now become part of the work-in-process inventory valuation process in the future. Bob, instead of directly answering Ted, picked up his telephone and called George, the corporate controller. George then requested that Ted visit him to go through his analysis and this recommendation.

Within thirty six hours, it was decided that the valuation of drilled holes was no longer appropriate given the significant change the new automated drilling machine had brought to the drilling operation within this PCB division. By not taking this action, work-in-process inventory valuation would have continued distorted. Beyond this decision, after speaking to the external auditors, George got permission to have Ted adjust the work-in-process inventory valuations for the last five months of the current business year to reflect this decision further. The proper adjusting journal entries were made, which then resulted in prior-period financial adjustments that restored pretax profit to just a bit more than 10%. The parent company dropped all consideration for closing or selling this division. Ted was a hero in Bob's eyes, not to mention all other PCB division employees whose jobs had been in jeopardy as well.

CHAPTER TWO

A Big Technological Change

Background Information

Another company sold its products made in China to a large chain of stores in the United States. This company's CEO opened this buying relationship with many small vendors in China long before President Nixon made his historic 1972 visit. This company's ownership shares traded on the New York Stock Exchange (NYSE). The various outside members of the board of directors urged the CEO to consider buying an Enterprise Resource Planning (ERP) system.

At first, the CEO resisted this proposal because of this software's high cost with the consulting services required to implement such a system. But over time, he relented and soon started an internal information technology (IT) project to seek an appropriate ERP software solution for his company. The project manager named to lead this endeavor was his executive vice president, who formed a small project team to perform a due diligence exercise to select an ERP software system. Because this company had not performed even one significant IT project in the last twenty years, an outside consulting firm was engaged to help them. The chosen firm was the same one they used as their outside audit firm.

The CEO was also the company's sole founder, dating back more than thirty five years. This man was a true entrepreneur, even from childhood. He started this company from his basement and soon sought a warehouse to store products as the business grew. He was very stingy about spending any of his money and went to great lengths to avoid doing so or, at the very least, minimizing having to do so.

This new ERP system project's budget exceeded 20 million dollars, which included new computer hardware and networking equipment, the ERP software, and consulting and training services. Although the company had ample means at its disposal to handle this large expenditure, it troubled the CEO. In this company's entire history, he had never spent this sum on any single purchase, including buildings he bought over the many years of being in business. He made it well known to his staff and to the consultants that this project had better deliver on all it promised, and then some, and in the prescribed project period.

The IT staff was the first group affected by this project. This department consisted of ten long-term, loyal employees; however, thanks to the attitude and past actions taken by the CEO, these people were at least twenty years behind the technology

curve involved in implementing the selected
ERP software. The CEO would not sanction any
employee, especially the IT staff, taking any outside
training classes, attending professional conferences,
or subscribing to professional journals and the like.
The reality of this unwritten company policy was
that he would not pay for people working for him to
learn anything new that they could subsequently
use to advance themselves by leaving this company
to take a better job opportunity. If they were to avail
themselves of such venues, it would have to be at
their expense as well as on their time. Thus, other
than a few IT subscriptions spread among this small
staff, no one ever received any technical training.
So now, given the need to implement this new ERP
system, an assortment of technical training classes
over that required for the new ERP solution was
needed. Training in databases, networking, and
client server technology was desperately needed
even before the ERP system could be successfully
installed and dealt with effectively.

Given this significant up front hurdle now facing
the CEO, he would not only have to pay suddenly for
all this required training, but also the added travel
expenses it entailed. As such, he decided to use
some space adjoining the ERP project "war" room to
have a classroom constructed so external instructors
would need to travel to the company, instead of
paying to have his entire IT staff travel to them.
In this manner, not only would he save money, but
also the IT staff would always be on site to deal with
any daily system support issues concerning their
existing and very old legacy systems that still ran
the business daily. He was still troubled that he
would have to fund all this training for the entire IT

staff in the latest and possibly greatest information technology and that these people might soon take this newfound knowledge elsewhere. However, the offset to this was that his business was about to benefit significantly from using this new ERP software system.

One day, as the company's project manager and consulting firm's project manager were walking down the executive office area hallway, the company project manager, Jennifer, turned to her counterpart, Arthur, and asked him the following question. "Once this new ERP system is in production, what percentage of the company's head count could be eliminated?" Without hesitation, Arthur answered, "50%." Jennifer nearly fell in shock at this news. When she regained some of her composure, she urged Arthur to explain his answer to her. He said that there at the corporate headquarters office, 350 people worked in various functional capacities ranging from sales and marketing to finance to procurement to legal and other administrative services, including human resources.

Arthur then told Jennifer to look around at what these people did, and more important, at how they performed their daily work in these various functions. He went on to explain that it was obvious to any casual observer that these employees were busy pushing many paper documents throughout the building. Paper was stacked in every direction. Arthur said that, in his many years of consulting, he had never encountered a business situation where so many people in one place at the same time operated in such a manner. With the advent of the new ERP software system, paper would become unnecessary and nearly extinct and, thus, would the need for so many paper-pushers.

Arthur's observation was accurate for a good reason. Again, it all stemmed from the CEO's attitude. As the business grew over the years, his solution to handling the added workload was to hire more clerical staff. Such people were plentiful in the local job market, and they were inexpensive. Further, his operating style did not allow much delegating to more senior managers, with the sole exception of the executive vice president who had been with him from nearly the beginning of this business enterprise. So, given the nature of most of the corporate head count and the online, real-time integrated nature of the new ERP software solution, Arthur's response to Jennifer's question was spot on, as shocking as this was to her. The most difficult part about this matter was that this workforce was largely very loyal and dedicated to this company. For the most part, they seemed to feel at home as among family working well together for a business providing products that made its customers feel so good to own or better yet to receive as gifts during some significant happy or sad life event. Jennifer knew that having to say goodbye to family would not be easy for any of them when that dreaded time, now not so far away, came.

Next Steps

Nearly seventy-five people soon staffed the ERP system project. Only ten were from the consulting firm with the remainder being company employees. This project's scope included the foundational business functions of sales and distribution, materials management, and financial management with all the necessary technical elements involved.

That most of the corporate headquarters office staff were clerical people was reflected in the composition of the company employees on the project team. The significance of this project was great given the lack of any real IT efforts during the last twenty years. So people selected to be involved in this endeavor knew how much would be expected of them, while they could potentially gain much in return if things went very well, including if the project was delivered on schedule and within its budget. They clearly understood that they had much to learn from their consultants given the wide and deep extent of this ERP system's functionality, the online and real-time integration of it all, and all the new technological features and functions involved with using it and getting it set up properly. It would require very long workdays of training by the consultants, countless hours of trial-and-error, hands-on implementation efforts, including thorough system testing, and a great deal of teamwork and patience by all parties. There were no project management disciplines, controls, methods, and tools to refer to from similar efforts, only what the consultants now provided. So this meant even more hours of explaining and subsequently having the company project team members understand the project management methods and tools to be used throughout this endeavor. Overall, this effort was huge in every way, meaning with all its technological, process (methods and tools), and people elements.

The idea of a project and a project team was foreign to this company from its lowest member on the organizational hierarchy up to the CEO. Thus, not all the challenges lay with just the company staff on the project team, but with all the consultants involved as well. This group was made of predominantly younger and thus less experienced consultants. They exhibited much energy, which compensated for whatever else they lacked, but their knowledge of this particular ERP software solution was extensive and without question. As is typical with such endeavors, everyone on the project team was fired up, as the effort finally got under way. A proper project kickoff event was conducted to explain to all the rules of engagement and the time line involved. Team-building activities were incorporated into the kickoff agenda and were held periodically after the kickoff meeting as well. The project was now launched, and it got under way in earnest.

The first milestone activity was to find out the state of all in-scope business processes. Because so much work was performed manually and was document-focused, written policies and procedures existed. These would need to be made current and then checked for accuracy and completeness. This effort took a bit of time to carry out, but it did not present itself at all as a big hurdle to anyone.

Once all these policies and procedures were collected and verified, their analysis got under way in earnest. The big challenge to the company staff on the project team now was to turn these many policies and procedures into improved business process procedures considered best business practices using the features and functions of the new ERP software system. This was when their new knowledge of this ERP system, acquired from their recent and many training classes, would be first tested. Further, this was when their many long work hours and its associated stress would begin to kick in as well.

Thus, it came as no surprise when some tensions began to flare. The consulting project manager, Arthur, expected this and was somewhat prepared to deal with it. Periodic team-building sessions were built into the project plan. Another device he started in this regard with Jennifer's concurrence was weekly, full-team, status review meetings.

Given the large size of the project team and the lack of physical space to gather everyone at once, the company cafeteria was soon determined the only place on site. Because the cafeteria was used to serve employees breakfast and lunch meals each workday, the end of the day was the only time this space was unoccupied. So every Wednesday at 6 p.m., the entire project team gathered in the cafeteria to hear from the co-project managers about the project's status and any open key issues and required decisions with key, upcoming, near-term activities.

A formal agenda was distributed before this meeting, but the meeting was conducted in as informal manner as possible to make everyone feel comfortable and more relaxed. Because the meeting time fell close to dinner and because the meeting place was in the company cafeteria, someone on the project team suggested to the co-project managers that if it were possible to do so, obtaining pizzas and sodas for everyone would be appreciated. So Jennifer and Arthur soon visited the executive vice president to seek his permission. He quickly agreed to this reasonable request and then went further by offering to pay for this expenditure from the project's already approved budget.

The following Wednesday evening, the entire project team came together for the weekly project status meeting, and this time, pizzas and sodas were provided to all present. The next day, the executive vice president summoned Jennifer and Arthur to his large office. He told them that he was getting several complaints about the pizzas and sodas being provided to the project team. The co-project managers didn't understand why the executive vice president even bothered to share this matter with them because the previous week, he approved their request.

He then explained to them that because this company had never done anything like this before for any employees, other staff not involved in this project and performing for this company in many other capacities over the years now thought it unfair to treat some to free pizzas and sodas while ignoring the other good employees. By telling this matter, the executive vice president was now trying to make Jennifer and Arthur feel guilty enough and sympathetic to his plight that they would agree to change their present course.

Unfortunately, for him, they felt the opposite, but it meant so much more coming from Jennifer as a company employee and manager. She told the executive vice president that she and her fellow company project team members were working far harder and were consistently putting in far more working hours than any other company staff members had in her 15 years with this company.

The least the company could do was feed these good people once a week at a modest cost to the company and to the project.

Although it was true that many other non-ERP project staff deserved some worthy recognition from this company, it would have to be up to the other senior managers, including the executive vice president, to address this matter with the CEO. Thus, she kicked the issue back to him with no sign of any regret. He wasn't very happy, but he realized that she was right because this matter rested with the CEO.

Meanwhile, and more important, the project progressed well. Considering how far behind the technology curve the IT department was and how clerical most company employees on the project team were, they acted open and very eager to learn now that the CEO enabled them to do so through this rare project opportunity. They put everything they had into this effort. Significant passion was shown by all; great learning and growth were achieved, and an intense appreciation was shown for the consultants driving the project.

As the company staff on the project team got farther along the project life cycle path and could better understand the need for and benefits of project management, they became disciples to all the other end users and employees in this company. They felt very strongly and favorably about the new ERP solution's features and functions, and even more important, about how things would be so much better in effectiveness once this system became operational. As they realized the power of the software's real-time integration and online information, they could understand, appreciate, and agree with Arthur's answer to Jennifer's question of sometime back about the upcoming impact this new system would have on employee head count at this company soon.

Resolution

The consultants were now highly regarded, given the situation shortly before. As previously mentioned, the consultants worked for the same firm as the company's outside auditors. The original team of consultants worked to gather all business requirements, and then they eventually helped this company's management select the particular ERP software solution now implemented.

Then they began to implement this system, but soon received considerable resistance from this company's management. The reason was simple. This group of consultants had ERP knowledge and experience, just not any with this vendor's software solution. As company management began to raise specific and detailed questions about the now-chosen ERP system, these consultants could not answer them well. Why this consulting firm undertook such a risky course of action to begin with was unclear.

The company executive vice president was so irritated he finally threatened to stop the project and dismiss all consultants. Only then, a consulting firm partner with knowledge and experience in using this ERP solution took the matter very seriously. He brought Arthur and visited the executive vice president with his senior managers to try to salvage the situation for this firm.

The consulting partner and Arthur carefully listened to them all vent their frustrations for a long while before they began to interject. By showing calm while giving precise and accurate answers to their many specific questions and concerns, the consulting partner and Arthur began to assure them. They requested that if these company senior managers would now allow their consulting firm to change the current consulting team with others more qualified, the project could forge ahead again. The executive vice president then spoke for them all by agreeing to let this consulting firm make amends, which they immediately did. A new consulting team was rapidly assembled with the proper knowledge and experience concerning this ERP software.

After nearly six months of intense and progressive efforts, the CEO now grew impatient with this entire endeavor. Though he was regularly briefed on the project status, he was still uncomfortable with it all, especially with all the money spent on this effort. Although his board of directors felt strongly about this ERP system and its need at this company, the CEO, whereas agreeing to allow the endeavor to go forward, never accepted its true value, especially given its high price tag.

He now decided to get the executive vice president and Jennifer together to share his unhappiness. He listened to them explain how well things were going from a schedule, delivery, knowledge transfer, and budget perspective. Nothing they said seemed to matter to the CEO. He then became unreasonable. He now wanted the project over much sooner than the remaining 4 months described and called for in the project plan. The CEO wanted the consultants gone even faster because they were largely the principal reason the project was so expensive now. He liked doing business in the manner he had done for the last 35 years and had no intention whatsoever of changing anything about that.

Jennifer now nobly tried to reason with him, but to no avail. At one point in their conversation, he went so far as to tell her that when her name appeared outside on the company sign, only then could she do whatever she pleased, but meanwhile, his name was on that sign, and he would call all the shots whether she liked it. She broke down and shortly left his office.

The executive vice president now took his turn at trying to calm the CEO and get him to be more reasonable about the project. At first, he, too, received the CEO's wrath. He decided to take it for the moment and to try again the next day, which he did.

After several more days of effort, the executive vice president emerged to report to Jennifer and Arthur that it was full speed as planned with no changes to be made to the project. Although they, and soon the entire project team, were relieved to hear this good news, it took a high toll on the executive vice president who shortly announced his early retirement from the company. He was tired of fighting against the CEO, and this latest battle would be his last. This scenario had played many times before about other business matters.

Anyway, because it was the ERP system project this time, all the people involved in it were disturbed by his news, but they understood why he sought to leave this company. Only then, the CEO decided to meet with his executive vice president to beg him to stay. The executive vice president was well prepared for this conversation, and soon thereafter, he emerged from it having obtained all that he asked for, including higher salary and stock options. Everyone was relieved to hear this good news and realized that he had even more influence than before, especially with the important project on which they were still working hard. Despite this episode, the endeavor moved forward with no obstacles.

Decisions were made rapidly; issues were handled quickly; and project communications were good. Not to say that things were perfect, but things were done as planned and well. The consultants acted professionally and conducted themselves well. They took seriously the crucial project task of knowledge transfer, and instead of waiting to near the end of the project, they performed this responsibility well daily to the benefit of all company employees on the project team. The CEO stayed out of the project's business while he received weekly status reports from Jennifer and Arthur as he had. Their working relationship was good, as it was with their project sponsor, the executive vice president.

As the project's end fast approached, everyone was re-energized for one last big push across the finish line that would make all their previous efforts finally come to fruition. It was a most exciting time for all these people and the company. The executive vice president was pleased with everything and everyone, and the CEO kept his distance at this crucial juncture. The project's spending was under

2 - A Big Technological Change

control and in line with the approved budget, thus the CEO could not complain. However, nothing anyone could do or say to him would ever make him comfortable with this very large expenditure, and they all knew this.

CHAPTER THREE

No Desired Technology Change

Background Information

Once, a manufacturing company was very successful and profitable. The CEO's father had started this company, but the current CEO was responsible for its immense growth in the last ten years or so. The product sold was in regular and repeated demand from very small to large retail business customers. As the company grew, so did all the associated paperwork. To maintain control, the current CEO developed a practice of reviewing and approving each document, such as all vendor payments, employee payroll and expense reimbursements, journal entries, purchase orders, and customer invoices, before it could be processed any further. When the business was much smaller, this was much more manageable, but now, it was a daily burden to him and everyone else relying on him. Even more, by now, he nonetheless could

45

not seem to break himself of this long and well-established habit. In addition, all this paperwork was processed manually. Because of this, the company staff was mainly clerical and, thus, used to operating this way. They knew a computer system would be far better for this company to have to handle the ever-expanding workload involved.

These employees were not alone in this belief. The outside auditors for the last several years had documented this recommendation in their annual management letter following the completion of their audit. And yearly, the CEO respectfully told the senior audit partner, "Thank you, but we will stay the course as is." His rationale was twofold. First, unless the external auditors had a more serious reason, such as a control shortcoming, there was no real or compelling reason for him to act on this nice suggestion. The other reason had to do with spending money. A new computer system would cost the company much money the CEO was not inclined to spend. In his mind, things were working just fine the way they were. Therefore, this scenario played over the last several years with the same result each time, to the frustration of the outside auditors and certain company staff and managers.

One day, following the issue of the latest audit report and management letter, the senior audit partner pleaded his case again to the company CEO. He got lucky this once, in that he must have caught the CEO on one of those days when the burden of handling all this paperwork had gotten the best of him. The CEO agreed to allow the senior audit partner to begin a study of their business processes and the potential need for a computerized

management information system. The senior audit
partner passed along this good news, not only to the
auditors directly involved with this client, but also,
more important, to this firm's consulting side. After
explaining the requirements to his counterpart,
the senior consulting partner, a consultant was
identified to do this work.

It turned out that the CEO and senior audit
partner had decided to have a lunch meeting to
discuss another more pressing matter having to do
with succession and estate planning. So, the audit
partner got the CEO to agree to a lunch date and
to invite the young consultant. The senior audit
partner very much wanted to get this consulting
study under way and on the right foot as soon
as possible by introducing this consultant to the
CEO before any work would commence. So, on
the designated day and time, the three met for
lunch at a restaurant nearby this company's office
headquarters. Beyond introducing the designated
consultant and a brief mention of the work he would
soon be addressing, the bulk of the lunch meeting
conversation centered on succession and estate
planning, as was intended.

The senior audit partner and young consultant
listened intently to the CEO describe his concerns
and intentions. Though this consultant, Daniel, was
by no means an expert in this subject, he carefully
listened to the CEO to understand him better. What
Daniel took away from this encounter was that the
CEO had been earning four million dollars a year
for the last several years. Earning even more money
was no longer his main motivation, but expanding
his business was. Further, being able to spend

much more time with his family, especially with his grandchildren, was now very important to the CEO. The time burden of having to review and approve all the paperwork in the office was his biggest obstacle to this desire. On the other hand, he felt the need to control the business in a manner that had worked so well for the last twenty-five years and for which he attributed part of the success of his business. The senior audit partner and CEO agreed to meet again in a few weeks to go over some ideas about succession and estate planning that the senior audit partner wanted to explore with the firm's senior tax partner and others. He promised to bring the CEO concrete suggestions to consider at their next meeting. So, the lunch adjourned and everyone went his separate way.

Daniel was finishing another consulting engagement and could not yet focus on this opportunity. Once that was completed, his first stop was to visit with the senior audit manager to gather details on this company from the audit work papers from the last few years. He took some copies and developed personal notes from this wealth of material before him. Daniel then had a few questions that he raised with members of the audit staff involved with this client to clarify certain matters. Once this first background effort had finished, he went to consider how best to tackle this important new assignment.

Next Steps

Daniel now carefully laid out a project plan of exactly what he felt was necessary to carry out his mission. The first milestone in his project plan involved interviewing all the company's middle managers to understand their business information requirements. By asking the auditors first to provide him with the names and functional job titles for each company manager, Daniel believed he now had the right cast of characters. He then allocated specific time for interviewing each manager, developed a questionnaire to use for these upcoming interview sessions, and worked with these business managers' administrative assistants to arrange a schedule to account for interviewing all these company managers. Now, he was ready to begin to interview these twenty-three people.

Daniel methodically worked through this information-gathering process with every manager. It took nearly three weeks to cycle through these twenty-three managers. He carefully and thoroughly gathered every piece of relevant and detailed information each manager needed to perform his or her job well. He collected samples of reports or their mocked-up notes for what each of these company managers needed. Daniel took copious notes that he sent to each manager after each interview session to verify for accuracy and completeness. Now, it was time to move into the next milestone phase of his consulting project plan.

Daniel needed to assemble all this information to analyze it. He previously determined the best way to go about doing so was to work back from the product of his assignment to deliver a written report of findings and recommendations and make a summary presentation of such to most company managers and the CEO. So, with the findings section of his forthcoming report, Daniel analyzed all the material he had collected by first seeking those things in common across most responses he obtained from these company managers as opposed to those things that were unique. Item by item, he painstakingly reviewed all this material and then documented the findings section of his upcoming final report. Most information requirements fell into the in common subsection instead of otherwise, which Daniel expected would happen.

Now, Daniel needed to develop the more challenging recommendations portion of his final report. He knew the CEO's negative feelings about acquiring a new computerized management information system and the auditor's desire for him to recommend such a solution. From the information he had gathered from all the company managers he interviewed, clearly, they, too, felt strongly about this matter and sided overwhelmingly with the external auditors for a new computerized management information system for this company. From all the materials Daniel collected and his subsequent analysis of it, he, too, agreed wholeheartedly with the company managers and auditors that a new computer system was called for without question. The issue now was how to present this case to the CEO in a manner that would result in his acquiescence of this recommendation by Daniel. He knew that he now had all the pertinent data to support this recommendation, but he also realized that data alone would not sway the CEO. It would take something far more compelling, but what.

Resolution

Daniel proceeded into the final phase of this consulting engagement's project plan, which had to do with both assembling and delivering his final report of findings and recommendations. He felt good about all the work he had done to date, in the support he received from the twenty-three company managers he had thoroughly interviewed, and in his ability to assemble a final, solid, detailed written report. His only uneasiness lay with his overriding recommendation about getting the company to initiate a proper search for a new computerized management information system and then to implement the selected vendor software solution. He knew that his data clearly supported this course of action, but he doubted seriously the ability of these data and his oratory and presentation skills alone to deliver this message so the CEO would be convinced accordingly. The summary presentation of his final report was now scheduled so most company managers, the CEO, and the senior audit partner could attend. Daniel's efforts were on schedule for developing all the materials for this final report of findings and recommendations, so he now moved to developing the summary Power Point version to be delivered at the big event in a few days.

As the short time leading up to the big show, when he alone would be in the spotlight, rapidly passed, Daniel was getting increasingly anxious about this thing. It had nothing to do with being nervous or unsure about any of the many details in his presentation, but about the outcome it would have on this company and its many dedicated hardworking managers in the future. He knew his presentation was fine, but it still lacked a compelling punch line to obtain the CEO's endorsement to move ahead with a search for a proper computerized management information system. Daniel racked his brain and reread his final report, with all the interview notes he had collected, many times, searching for the right words to win over the CEO to gain his acceptance of his overriding recommendation.

Now, it was thirty hours from show time, and still, Daniel could not determine what else he could do to rectify this matter. He did not aspire to be just another failed effort by his consulting firm with this CEO concerning this important matter. Daniel wanted to deliver for the auditors and company managers who now depended on him. He wanted to be persuasive enough to overcome any resistance the CEO could muster. The rest of the day passed, and still, Daniel lacked the punch line he sought to deliver the next morning. That night, while lying awake in bed, it finally came to him. He arose early the next morning to put the final touches to his summary presentation.

When the clock struck the appointed hour, Daniel was ready. Everyone was assembled in the large conference room at the company headquarters office building to hear from Daniel. At the last minute, he decided not to hand out copies of his upcoming presentation until he had finished delivering it all, which was not his normal practice under such circumstances. He wanted everyone, and especially the CEO, to focus on all the many details in his presentation's content as he delivered it, without jumping ahead of him. Daniel constructed this presentation to have it build to a climactic final moment.

He now diligently delivered this report. Daniel first explained to the audience the approach he took in performing this study, and then he named every company manager he had interviewed. Next, he explained each of his findings, starting with those he found in common for most company managers, compared with those that were somewhat different. Daniel now turned everyone's attention to the recommendations portion of his presentation.

To this point in his presentation, the audience raised very few questions other than to obtain certain clarifications occasionally from Daniel. Everyone's level of expectation was now high as the presentation turned its focus on Daniel's forthcoming recommendations. He began this portion of his presentation with some small and obvious suggestions for improvement with which no one could disagree. This generated some conversation, mainly about next steps in being able to implement these good recommendations. Daniel now began to proceed to his more substantial recommendations. He touched on a few of these that clearly headed toward his overriding recommendation pertaining to a new computerized management information system. Daniel finished his formal remarks as the screen displaying his slides went blank. He then deliberately turned to face the CEO directly to address him with his next comments.

Daniel respectfully reminded the CEO of two things. The first had to do with replaying the CEO's words to him that Daniel had heard at that introductory lunch meeting with the CEO and senior audit partner eight weeks earlier. These remarks had to do with the CEO complaining about all the time he spent reviewing and approving documents, which stole the precious time he preferred to spend now with his family, especially his grandchildren. The CEO nodded that he had made such comments.

The second comment Daniel made had to do with something he learned from several company managers while he was performing his due-diligence, information-gathering process. What he now told the audience was that several more senior managers were new in their various positions at this company. The CEO had decided about a year ago to bring these people into the company to help him manage the more recent growth in business that had already occurred and that was expected to continue, looking ahead. Besides hiring these more senior managers, the CEO had also decided to provide them with financial incentives over their base salaries. Because this was a privately held company, he decided to create a profit-sharing program for the specific benefit of these people and their direct reports. Daniel then reminded the CEO that he was no longer motivated to acquire more money, but as the company expanded, these senior managers could earn much more compensation because of this new profit-sharing arrangement the CEO had designed for them.

He delivered his presentation's punch line. Daniel said to the CEO that obtaining a computerized management information system was not specifically intended for his use and benefit, but it was instead for these company managers. For them to perform their various jobs well and help the company handle successfully the expected additional growth in business, these good people required a new computer system.

Everyone, including Daniel, now eagerly waited
for the CEO to respond to this final overriding
recommendation. When he did, he told them he had
been running this business all these many years
without the benefit of such a tool. However, if that
was what they needed to do their jobs well, to help
expand the company, and to maximize their bonuses,
then he would allow them to proceed with a search
for a new computerized management information
system. The audience chimed in that such a tool
would enable them and their staffs to be more
effective, and timelier, more accurate, and more
complete information would help them greatly.

Therefore, Daniel changed the CEO's mind
by putting the emphasis for using the new
computerized management information system
on his staff, and not on him, because he did not
intend ever to touch it. For thirty-five years, he
had run this business without a computer system,
and he would not start now. However, in support
of his senior staff, he would permit it. Daniel and
his consulting firm shortly got a search process
for a new computerized management information
system under way by working with the CEO's senior
management staff. Over time, a new computer
system was selected and then implemented for the
benefit of this entire manufacturing company and its
many senior and middle managers.

Leading Project-Driven Change

CHAPTER FOUR

A Downsized
Technology Change

Background Information

A manufacturing company had a very long corporate history. This company dated back more than 125 years. For anyone fortunate enough to visit it and the surrounding town in the mountains, it was an interesting historic experience to glimpse into part of America's industrial past. The company in more recent years had declined substantially from its former glory days. It also went through many ownership changes that first took it from a private to a public company and now, more recently, back to a private business enterprise. The new owners were former senior managers at the public corporation from which they recently acquired it. They believed that by addressing certain key business matters, they could turn around this old company's fortunes while benefiting financially themselves.

The sole product this company produced came from a natural resource with stable demand from many other manufacturing businesses that required this material in their manufacturing processes to produce their products for sale. This company mined this natural resource and then refined it, making it purer before transferring it to outside customers. The company's location was well off the beaten path, so in time, a small town sprung up around it. Almost every other business in this town, including the local railroad and water utility, existed to support this company. The company owned about everything in sight except a local restaurant and barbershop. Thus, as the company's fortunes rose over time, so, too, did those of the town, and vice versa. In time, the company owned all the homes in which their employees and their families lived. In its time, the business and town had thrived, but no longer.

It was a slow, but steady, decline over the last twenty years that brought things to their state of affairs. The plant controller now regularly complained to the new owners that because of the many waves of staff reductions over the last ten years by the various former company owners, his current small financial staff could no longer adequately maintain the highly sophisticated cost accounting system. It had been first put into service twenty-five years earlier by a well-regarded management consultant.

This management consultant then was a nationally renowned cost accounting expert who had worked for one of the leading audit and consulting firms. He was engaged at that time to help this company with the internal development of a complex cost accounting system, given the nature of all the plant operations needing to be tracked and controlled. During this period, the company was thriving and growing, thus, management's information requirements were as well. This consultant performed a thorough plant operations study and analysis before designing this cost accounting system for some external computer programmers to develop fully. This system was the latest and produced countless hardcopy reports daily with detailed operational and financial information. No one had the necessary knowledge and time or any inclination to challenge this highly regarded system although the business was drastically curtailed in the many intervening years.

The new management team now decided to engage this consulting firm to commence a study of this entire situation. Now, the management consultant with all the cost accounting knowledge and experience who had created this complex system was no longer an employee of this firm. He had retired long ago, and it was believed that he had died. So, the consulting firm did the next-best thing, which was to identify a modern management consultant with all the requisite cost accounting background to step in. Henry was shortly sent to this company's headquarters office to begin this work. His given mission meant either to find a way to reduce the burden of maintaining this sophisticated cost accounting system on the controller and his small financial staff, or to replace it with something much simpler, much better suited for present operations, and much less expensive.

Next Steps

Henry had the advantage of having full access to all the recent audit work papers because his consulting firm also represented this company as their auditors for many years. He began to review the extensive audit work papers to prepare for his new consulting assignment. Henry spoke to the audit partner to determine more recent information having to do with the latest ownership change. He specifically wanted to know who these new owners were and what their functional backgrounds were. He learned that one owner had significant sales and marketing management knowledge and experience, another owner had an extensive operations management background, and the third owner had a long financial management career. Henry next developed a project plan for tackling this effort, and he arranged to visit the company by first calling on the company controller, Scott.

Soon, Henry was in Scott's small office at the company headquarters, gathering additional information and answers to the many questions he had prepared for this encounter. From this long conversation, Henry returned to his office to finalize his project plan and approach for performing this upcoming study and analysis. When he completed this work, he sent it on to Scott for final verification. After Scott's concurrence, he was now ready to begin the real consulting work, beginning with the first milestone in his project plan. Henry was to get a thorough plant tour and then start to interview all company middle managers. This was to be followed by interviewing the new company owners.

Henry obtained an extensive plant tour from
an operational supervisor, Matthew, who had
worked for this company over the last 30 years.
He spent an entire day following Matthew around
the sprawling manufacturing facility as they went
through every operation in detail, covering the entire
manufacturing process. Henry raised countless
questions for Matthew in each work center they
visited. Henry took detailed notes throughout the
effort. He found two specific things interesting
during this tour.

The first of these observations he now raised
with Matthew at the end of the day. It had to do with
physical inventory. This product, at times, would
accumulate beyond the physical limits for which
indoor storage space could provide. Thus, the excess
work-in-process inventory was stored outside. It
appeared as a pile of dirty sand and, being outside,
was subjected to the elements of nature. This was
further compounded by this company's location
located high in the mountains where it tended
to be windy. Henry wondered how the company
could conduct a proper physical inventory of what
appeared a pile of dirt blowing around outside in the
wind. Matthew went on to explain how this situation
was handled during a physical inventory count, but
Henry still thought it was odd. Then, if the auditors
were satisfied with this procedure, so would he.
Henry had much bigger things to consider instead of
this matter.

The other observation Henry had from his tour
from Matthew was that throughout this plant he
saw supervisors of work centers with very few
workers in them. At this time, he decided not to

question Matthew about this observation, but kept it to himself for some not yet determined future use on his part.

Anyway, Henry finished his detailed notes from this plant tour and sent them to Matthew for his final verification. Matthew shortly got back to Henry, indicating that his extensive notes were accurate and complete. Meanwhile, while he awaited Matthew's response, Henry began to develop a questionnaire that he planned to use in his upcoming middle manager interviews. He already had determined a list of people to interview, and it numbered eighteen company managers. Henry now worked on arranging a schedule for interviewing each of these managers over the next few weeks.

He now began to conduct these interviews with the predetermined group of company middle managers. After he held the first few interviews, Henry decided to make some minor modifications in his questionnaire, and he became aware of three more managers with whom he needed to speak. He made all necessary adjustments and continued these interview sessions daily. After every interview session, Henry provided time in his schedule to document his detailed notes fully, so he could send them back to the managers he had just interviewed for their review and validation for accuracy and completeness. Most of the time, they responded to him that his notes were perfect, and on occasion, someone suggested a minor clarification, mainly because of Henry's wording.

Henry moved forward with the next stage of his consulting activities, which had to do with thoroughly analyzing all the details he had collected to date from his facility tour and all the interviews with the company managers. He took the approach in his analysis to focus primarily on those matters pertinent to his assignment's main purpose, which had to do with examining how the cost accounting system could be downsized to fit modern business circumstances. Only after he dealt with that crucial subject would he provide any additional recommendations to the new company owners. Henry realized that he needed to spend some time with the controller, Scott, and his cost accounting manager, Paul, to understand the existing cost accounting system better.

Henry arranged to spend time with both Scott and Paul to understand thoroughly all the many inputs to this sophisticated cost accounting system and all its many daily outputs. Henry already had a detailed flowchart of how the cost accounting system processed these inputs to produce its various outputs in operational and financial hardcopy reports.

Henry found a document among the audit work papers dating back twenty-five years to its creator, the cost accounting expert from his consulting firm. At first glance, it seemed very complicated, but supplemental detailed notes explained it well. To Henry, as another cost accounting expert, it was a piece of beautiful artwork. After spending much time with both Scott and Paul, Henry was now ready to return to his office to complete his analysis.

Resolution

Henry carefully examined every detail of information he had. It was not easy to determine exactly how to downsize this labor-intensive cost accounting system; on the other hand, he had no desire to replace it unless there was no possible alternative. As Henry came to appreciate right then, it was far easier and much more fun to create a cost accounting system or even enhance one than to change or reduce one to fit a different set of, and in this case, much worse, business circumstances. He struggled somewhat with this part of his assignment, but given more time and further effort, he found several ways to scale back the existing cost accounting system this company used. Henry eventually figured this all out by working backward from what information was required by the current set of company middle managers. By doing so, he was able to develop recommendations to eliminate the collection of certain data inputs, while also cutting out many output hardcopy reports.

According to his work plan, Henry was to interview each of the three owners separately at this point in the project. The main purpose in doing so was not to gain any details needed to carry out his crucial mission, but to meet and to know better each of them while telling his activities to date and progress. Further, he wanted to see if they had any unique information requirements that were more than those of the middle managers with whom he had already met. Finally, Henry would soon need to complete his final report and deliver a summary presentation to these three owners with some of their key direct reports, so getting to know more about these owners was another important objective of his.

Each of these conversations was scheduled for thirty minutes. The first half of this time was for

Henry to provide them with his status update, and
the balance of the interview time was for him to
gain any other relevant information they might have
to offer. He soon met with the first of these three
new owners, the sales and marketing management
leader for the company. Nothing special about this
encounter added to Henry's remaining planned work
tasks. Things seemed to go well, and this owner
seemed satisfied with what he heard from Henry.

Henry moved to the next interview with the
owner with operations management responsibility
for this company. This discussion was more
interesting to Henry because a significant focus
of his work to date had been in this business'
operations and financial areas. This owner asked
Henry many detailed questions about whom he had
met with and about exactly what he had done to
date. Then, this owner told Henry he had no other
management information requirements beyond
those of his direct reports. His only concern was that
they all had the information they needed and on a
timely basis to perform their jobs well.

Finally, Henry held his last interview with the
owner with the extensive financial management
background. As before, Henry carefully explained
his various activities to date before turning his
attention to seeking this owner's specific information
requirements. This person told Henry that in his
previous role for the former company who had
owned this business, he had much exposure several
times to the cost accounting system now in use. He
thought that it was too complicated and expensive
to maintain, and he would favor getting rid of it and
finding a more suitable and less costly replacement

system. Further, he said that the owner with all the operations management knowledge and experience was unfamiliar with the many inputs and outputs produced by the existing cost accounting system. Although he did not shut the door to downsizing the current cost accounting system, he let Henry know where he stood on this important matter.

Henry had taken very good notes from his conversations with these three owners, which he incorporated into his final analysis and preparation work for his upcoming report of findings and recommendations. Besides this final written detailed report, he developed a summary version to use as material for his upcoming presentation to these owners and select middle managers. Henry took his time in deliberating about what to include in these documents. He considered everything at his disposal and whether he still required something more.

Finally, the time came for him to finish his work on these two documents, but before these two documents could be issued to the client, an internal consulting firm review was required. Henry's manager was the consulting partner in charge of his office, Edward. He read all Henry's material and then raised one objection. In both these documents, Henry stated a finding pertaining to an early observation he made that had to do with the ratio of plant workers to supervisors. Henry had observed throughout the plant that a supervisor frequently managed only one, two, or no more than three workers. Although he had no other facts to base this on, his many years of management consulting experience, operating in many other manufacturing environments, told him this was a problem worth noting.

Edward told Henry that although his observation had merit, unless he could back it up with facts, he would not allow Henry to put this finding in either of these two documents. Henry considered this matter a challenge, and he was determined to find the facts to support his contention. He began to focus on exactly how he would carry out this objective. Fortunately, he did not require much time, nor did he have to look very far. In his consulting firm's office, there was a small library. In this library, he discovered a very old book having to do with the kinds of cost accounting systems used in specific industries. A chapter in this book had to do with the industry in which this company operated. In this chapter, he discovered the name of an expert who had written this section of the book. Henry sought to contact this expert at a university in another state to discuss his observation, with the goals of vindicating himself and to satisfy Edward's objection. He finally contacted this professor, and they discussed the matter on Henry's mind. This professor agreed with Henry that the ratio of workers to supervisors, instead of being two or three to one, should be more like six or seven to one for this industry. Only then did Edward allow Henry to quote this authority in both his reports.

It was time to deliver his summary presentation. First, Henry told the audience the approach he used to collect his information, and he mentioned the name and functional title of each manager with whom he had spoken. He thanked Matthew for all his time and efforts in providing him with

a thorough plant tour. Then, he went through
all his various findings before addressing the
recommendations section of his presentation.
He told his detailed suggestions for reducing the
number of inputs to and outputs from the existing
cost accounting system, thus eliminating the need
to replace it. He further put forth estimates on how
many man-hours this would save the company,
especially the controller's staff, monthly in the
future. Instead of turning off parts of the current
cost accounting system, they would no longer be
used. The effect would be the same as downsizing
the system.

All the people seemed fine with this conclusion,
and they seemed pleased with this result. Finally,
Henry got to the last section of his presentation,
which he added at the last minute. It had to do with
his observations on things management should
consider addressing soon. The main item had to
do with the ratio of workers to supervisors. After
listening to Henry, both the operations and financial
management owners chimed in. They were both
astounded by Henry's observation. They did not
doubt for an instant the statement coming from the
university professor about this ratio needing to be
much higher. Matthew spoke up saying that over
the last 10 years of worker cutbacks, no one had
considered the number of supervisors in the plant,
yet alone the ratio of workers to supervisors. Various
parties had done this process very gradually, thus,
it got under the radar screens of operational and
financial management, both under the prior and the
current ownership. At this point, all three owners
thanked Henry for his fine efforts and many good
recommendations, and they promised to address

his observation having to do with the ratio of
workers to supervisors immediately. They were very
grateful to him for having discovered and reported
the important matter that had been overlooked
internally for so long to them.

CHAPTER FIVE

No Focus on a Technology Change

Background Information

A long-standing company was in the business of manufacturing printed packaging materials to a variety of commercial customers. A man who later passed it along to his five adult children after his death started this business. The makeup of these five siblings was four men and one woman. The four brothers were involved in different aspects of the business daily. The company was privately held and equally owned by all five siblings. The company's products were very reasonably priced, provided at high quality, and produced in more than enough quantities to satisfy its disparate base of business customers. Many size and printing differences coming from these customer requirements made for most of the business challenge over the years; however, something more was happening here that was not so obvious from the outside looking in.

The four brothers occupied executive positions
in the company. Each fulfilled a particular
business function such as sales and marketing
management, operational management, finance and
administration management, and so on. Whenever
an enterprise-wide decision was called for, they
could not agree. As a result, over time, each brother
had several line managers directly reporting to
him who took the same position on every issue as
each brother took on specific business matters,
thus putting them in direct conflict with all the
others. What you had was one business run by four
kings and their separate armies, all pulling in the
opposite direction from one another. Simultaneously,
a new need surfaced having to do with acquiring
a standard cost accounting system. This business
requirement was treated as an exception in that
three of the four brothers agreed with this need,
whereas the remaining brother did not seem to care
either way.

The next step was to request a management
consultant from the same firm that provided this
company with its audit and tax services to address
this important new business requirement. To
determine the best standard cost accounting system
vendor solution, they realized that all relevant
and detailed business requirements supporting
this contention needed to be gathered and then
analyzed. This was also acceptable to them because
this company had never analyzed their business
processes in any detail before. Soon, a management
consultant was identified for this endeavor. He was
then sent to this company's headquarters to begin
this important work. Brian soon met with one of
the brothers, an executive responsible for finance

and administrative matters at this company. This executive explained to Brian the rationale behind this newly identified business need for a standard cost accounting system. Brian listened carefully and attentively to him, all the while taking detailed notes of their conversation. In addition, this executive told much information to Brian pertaining to the company's long history. Brian found this all fascinating.

One of the many things Brian learned from his conversation with the finance and administration executive was that the company had been losing money for the last several years and had borrowed many funds during this time from a local bank the company had been doing business with since the original owner was alive and in charge of things. The bank recently decided not to allow an increase in the company's line of credit as they had in years past unless the company's executives agreed to take certain actions and to do so right away. One of these requirements had to do with this company acquiring a standard cost accounting system. The source of this request originated with the outside auditors who determined this need a few years ago. Although most of the four brothers had no particular objection to this request now coming from their bank, they all knew the effort to achieve this goal would cost them some time and money. Despite this fact, they had already agreed to comply with the bank's request, which was why Brian and his management consulting firm were now called on to address this need.

Among the other requirements put forth by their bank was one that would be far more difficult to achieve than many others would. This one had to do with getting the four brothers to work in concert for this business' betterment. Although none of the brothers would publicly disagree to satisfy their bank's request, none of them considered how he would achieve such a harmonious result. These four brothers had been operating in this manner for so many years that they did not even know how to work any other way. Fulfilling this requirement by their bank would be their greatest challenge, and two of them wondered how the bank could measure improvement, even if they could figure out how to achieve what was asked of them all.

Next Steps

Brian mapped his approach for conducting this consulting engagement. He determined that he first needed to understand the business operating environment and then gather detailed information requirements from senior and middle-level company managers. To this end, he sought someone qualified to provide him with a thorough plant tour. Brian also made various inquiries about whom he needed to interview to gather all the management information requirements. He first conferred with this client's senior audit manager to get his input. Then, he checked in briefly with the financial and administrative executive to seek his verification of this list of managers to interview. This executive suggested to Brian that he reach out to his brother, the operations executive. Brian soon did this and obtained the names of three more managers to interview. Once he had all this information, Brian was able to begin to develop a project plan for this important assignment.

Brain began to execute his work plan. The first major activity was to obtain a detailed plant tour. His tour guide was the plant manager, Kenneth, who reported directly to the operations executive. Brian and Kenneth spent most of an entire day going through each work center in detail. Brian asked many questions and took thorough notes. The next day, he sent a clean copy of these notes to Kenneth for his verification. Kenneth shortly replied that Brian's detailed notes were accurate and complete.

Brian moved to his next significant project activity to interview sixteen senior and mid-level company managers. These sessions happened over two and a half weeks in no particular order other than when these managers were available to Brian for this purpose. Each interview required at least an hour, and some went for as long as two and a half hours. The sessions with each of the four brother executives took the shortest time, because for the most part, they deferred to their subordinates. Everyone was cooperative, and almost all supported Brian's efforts for the company as a whole. Again, Brian took detailed notes at each session and, afterward, sent these notes to every interviewee to have them checked for accuracy and completeness.

Once this process had finished, Brian turned his attention to analyzing all the information he had collected to date. He now considered what the product of his consulting work would need to be, and he decided to share his findings and recommendations. So, as Brian analyzed all the material before him, he began to sort all the details into one of these primary two sections of his final report. The items going into the findings portion were far easier to handle and required much less of his time. On the other hand, his recommendations were much more challenging.

As he considered each recommendation, Brian tried to imagine the resistance he might encounter from any company manager. Most seemed to Brian to agree that acquiring a standard cost accounting system was a worthy idea from which they, as well as the entire company, would benefit. However, Brian was by now well aware that when these managers were assembled and had to declare their viewpoints publicly and support of one of his recommendations, they would ultimately side with the position of the executive brother to whom each reported. Thus, for Brian, this was the challenge of this management consulting engagement.

To begin with, the idea for acquiring a standard cost accounting system came from this company's bank. The financial and administrative executive had no choice but to accept the bank's recommendation, especially given how much credit the bank had already extended to the company to date. Neither he, nor any of his direct reports, thought this recommendation was that important and urgent for him to address now. As a result, they were all not very interested in the bank's recommendation, but considering the source and present circumstances in which the company found itself, they reluctantly had to play along. They all contributed solid information to Brian about their requirements and thoughts about the value of having a functioning standard cost accounting system. Some of these managers were much more familiar with using and having the information a standard cost accounting system could provide from their prior places of employment, whereas other managers were not.

Resolution

Brian developed his final report of findings and recommendations, especially concentrating on the recommendations section. He was readily able to explain the value of having and using a standard cost accounting system. Brian described all the specific data inputs that would need to be gathered and subsequently fed into this system, as well as all the information that such a system could provide to all the company managers. Nothing was special or unique about this manufacturing company from all the many others Brian had worked with previously in his roles as a management consultant and as a corporate cost accounting manager.

Brian carefully chose the words he used in his detailed final report to company management. Further, earlier in this engagement, the financial and administrative executive asked him not only to produce a detailed final written report but also to prepare and then deliver a summary version as a presentation to his fellow brother executives and their managers whom they wished to bring along to this event. So, Brian was now busy finishing these two important documents. The presentation was already scheduled and was now just a few days away from happening.

Brian rapidly, but methodically, proceeded with
his work on completing these documents. He reread
his detailed final report and felt confident about
it. Then, he focused on his summary presentation.
Here, too, he felt good about all he had in it, but
he still wasn't as sure about the recommendation
section of this material. It wasn't so much that
anything was missing or was not said properly,
but that his recommendations lacked any sense
of urgency except to satisfy the company's bank.
Fortunately for him, things were about to change
dramatically and for the better.

All the while, as Brian was busy consulting,
the financial and administrative executive with his
three brothers were constantly communicating and
negotiating with their bank's senior lending officers.
Aside from the bank's recommendation about this
company acquiring and using a standard cost
accounting system, they had several other important
matters about which they were far more insistent.
These four brother executives had requested their
bank substantially increase the company's current
credit line limit. The company's fortunes in sales
and profitability were now in a steep decline, and
they desperately needed more cash to stay afloat
and pay their suppliers, not to mention all their
employees. The bank's patience with these four
brothers was wearing very thin. This bank had been
dealing with them all for a very long time, even
when their father was still living and running this
company he had started. The bank's representatives
had been promised many things during these many
intervening years by these four brothers only to see
most of their promises never happen. The financial
and administrative executive was the oldest brother

now in his early sixties, followed by the sales and marketing executive in his late fifties, followed by the operations executive in his mid-fifties, and finally followed by the facilities executive in his early fifties. They all had worked full time in this company business since they had graduated from college.

The bank representatives after several difficult meetings with these four brothers were now prepared to declare the bank's intentions about the open request to increase the company's credit line limit. One key requirement the bank put to these four brothers, and for that matter had made repeatedly to them over the years since their father died, was that they needed to work together instead of tearing the company apart. These four brothers agreed to act in this manner whenever the bank representatives brought up this subject to them, but they never did so. This time, the bank decided to use this issue as their deal breaker. Unless these four brothers could quickly show the bank, by producing tangible proof of such cooperation, the request to increase the company's credit line limit would be completely denied. The bank further threatened to act to change the terms and conditions of the existing credit line agreement, effectively putting even more pressure on this company and these four executive brothers. The financial and administrative executive brother knew, given the company's current weak financial condition, that no other lending institution would help them. He had done a bit of due diligence by testing the waters, and by now, he had enough proof of this contention, which he had already shared with his brothers.

It was at that most crucial moment that the uninvolved sibling, their sister, suddenly became very much involved. The bank's representatives had already contacted her about this situation because she was an equal owner of the business like her four brothers. She had called an urgent meeting at the company's headquarters with her four brothers to discuss the bank's highly likely forthcoming decision to reject their request to increase the company's existing credit line limit. She took complete charge of this meeting and told her four brothers, because she was the oldest, that they should all be ashamed for allowing the company their father had created and passed on to them to land in such a dire situation. She went on to berate them repeatedly for their bad behavior over the years that was now tearing the company apart. She further told them they were jeopardizing her financial situation, because she was now retired from her teaching career.

The brothers all pledged to their older sister that they would immediately cooperate for the betterment of the overall interests of the business instead of their separate interests. They would prove this immediately and tangibly to her and over time to the bank. They then asked her on their behalf to meet with the bank's representatives to try to get them to reconsider their position and allow the company to obtain some additional cash to rectify things within a strict ninety-day period. Only then did she agree to do so.

Two days following this big sibling meeting, it was time for Brian to make his summary presentation to the four executive brothers with their invited subordinate managers. He first went over the findings section with hardly any feedback from the audience gathered before him. Then, he began to address the recommendations part of his presentation. As he did so, he noticed that each of the four executive brothers was taking an active role in speaking out and in unison about accepting and implementing each recommendation Brian put forth. Not only did Brian notice this happening, but also more important, the invited subordinate managers readily noticed this different atmosphere regarding both the attitude and behavior coming from all four of these executives. None of them had ever witnessed such unity and common purpose displayed in public before, coming from these four brothers. Each recommendation was accepted as is, and the discussions centered on when and how to get them put in place, including selecting a vendor solution for a standard cost accounting system. Although Brian was leading this presentation, he knew that something else much more powerful than he and his presentation was at work here, and he didn't seem to mind one bit.

CHAPTER SIX

Lessons Learned

T he five short stories just presented share various elements that are in common and others that are not. By analyzing the positive and not so positive aspects of these five tales and the learning their outcomes provide us, we can better see project leadership in operation for what it is or is not.

In the first short story titled "A Bit of a Technology Change," we have a challenging business situation. Because of the recent large expenditure for a computerized drilling machine and the onset of wild fluctuations in its monthly financial results, the PCB division is under severe duress from its parent corporation. One controller has already been demoted, while his replacement has limited time to rectify matters. No obvious signs offer any good clues to help solve the challenge. It took some time, much effort, and significant focus and persistence to discover the source of the problem. Everyone knew about the new computerized drilling machine replacing the former manual drill press, yet no one ever suspected the impact it would and did have on work-in-process inventory and profitability for such a long time. The turmoil it created was unnecessary and painful. In the end, a positive result was finally obtained.

Then, in the second short story titled "A Big Technology Change," we have a CEO with great pride in his achievements in creating and expanding this business enterprise over a long period. However, he never accepted the notion that his company required an ERP system. His handpicked board of directors and many of his staff did, but not he. Yes, he allowed the project to move forward, but its high cost troubled him to no end. Despite this and that the consulting firm's initial team was tossed out and then allowed to be replaced, the project achieved all its intended business objectives. The company environment was much further behind the technology curve than most others of its time and size were, and thus had a far more significant hurdle to overcome. There was no previous project management knowledge and experience in this company before this endeavor. However, thanks to a strong and determined project sponsor, the company's executive vice president, and to the great efforts put forth by the entire project team, including the consultants, this project was successful. In this case, the determination of the many overcame the resistance of one, who happened to be the company's CEO.

In the third short story titled "No Desired Technology Change," we see another CEO resisting change. He associated the need for a computerized management information system, expressed by

his company's external auditors, as something
he would need to use and benefit from. Given his
long-standing practice of reviewing every financial
transaction, he saw no reason to spend any money
on such an unnecessary solution. Despite his
resistance, neither the consultant and the auditors
nor the company's senior managers let this deter
them from persisting with this important initiative.
When the onus was placed on his senior managers
for using and benefiting from a computerized
management information system, this CEO relented.
The CEO was a reasonable business executive who
cared enough to provide his senior managers with
a profit-sharing incentive program. The consultant
used this sensitivity to get the CEO to agree to allow
his senior managers to begin a search process for a
new computerized management information system
for their use in this company.

Next, in the fourth short story, titled "A
Downsized Technology Change," we witness a
unique situation and set of circumstances. A series
of ownership changes hid an important business
and operational problem. The consultant was tasked
with another mission and objective; however, he
discovered this matter as a by-product of his due
diligence efforts. Then, his consulting manager
presented this consultant with another hurdle
to overcome. The consultant's persistence, not
to mention a bit of good fortune, enabled him to
meet this challenge to share his discovery with the
company's new owners. In addition, the main goal
of downsizing the existing cost accounting system
was achieved without the necessity of replacing it
or dramatically altering it. Instead, the consultant's
creativity resulted in reducing the many data inputs

and thus the outputs without having to overhaul the current system. This effectively achieved the stated aim of downsizing the current cost accounting system to relieve the burden it placed on the small financial staff that had to maintain it in the future.

Finally, in the fifth short story, titled "No Focus on a Technology Change," we have a terrible business scenario. One company is run by four chiefs who prefer pulling the company in four directions, rather than working together for a united purpose. Effectively, they were pulling the company apart by their sibling rivalry decisions and actions. It took one external force, the company's bank, and another outside party, their sister who had equal ownership interest in this company but had no daily operational responsibilities, to get the full attention of these four executive brothers. By nearly driving the company to the financial brink because of their past and more recent decisions and actions, these four brothers were forced to obey their sister to obtain the bank's salvation. This was expressed in a financial lifeline through an increase in the company's line of credit limit and the additional time given for them to change their bad behavior. One of the first ways these four brother executives could show this was in their agreement to obtain and then use a new standard cost accounting system for the direct benefit of the company and its middle managers in particular.

As we reflect on these five short stories, we should first recognize that they resulted in positive outcomes. Despite the various and many challenges the key participants, internal or external, encountered, in the end, their determination overcame any resistance placed before them, even when it was from the company CEO. Win-win results were achieved for the companies involved in these five short stories, the executives and managers of these companies, and the person responsible for driving the required efforts to make it happen. In four out of five situations, this person was an outside consultant, while in one case, it involved an inside company manager. Four of the five short stories were about manufacturing companies operating in different industries, while one was not.

In two of these five short stories, the resistance coming from the particular CEO was long standing and strong. Even after consulting efforts were well under way, these CEOs persisted in their specific objections. In the first of these short stories, the computerized drilling machine's unintended consequences almost caused the parent corporation to sell or shut down its PCB division. In the fourth short story, the unintended discovery of the worker-to-supervisor ratio proved almost as important in the end as was the engagement's primary purpose, the high-maintenance effort involved in supporting the existing cost accounting system. Finally, in the fifth short story, the lack of any interest by the four executive brothers in acting in harmony, for the sake of the company overall and for the sake of their own staffs, nearly destroyed this business enterprise started many years earlier by their now dead father.

What we learn from reviewing these five short stories are the following lessons:

- From the first short story, little things can make a significant difference, and the obvious culprit, in this case, the computerized drilling machine and its associated impact on work-in-process inventory valuation, was overlooked for a long while.

- From the second short story, despite being so far behind the learning curve in information technology and project management knowledge and experience and the strong resistance of the CEO, the project became a success. Clearly, the need for an ERP system was well justified. In the end, the will of the many overcame the power of one, the CEO.

- From the third short story, the onus for the use of a computerized management information system was taken off the CEO. Here again, the need for such a solution was well justified. By placing the need for and benefit of the computerized management information system on the company's senior managers, the CEO finally relented and allowed them to proceed.

- From the fourth short story, it took the very astute observation of an outsider to spot and then report a situation missed daily by many insiders for a long period. Further, the first and early response pertaining to replacing the current cost accounting system proved not the best solution for this company in the end.

In short, these five short stories are designed to share the following traits about *Project-Driven Change Leadership*:

- Focusing on results
- Determination
- Persistence
- Professionally challenging the status quo
- Performing due diligence
- Overcoming obstacles
- Seeking opportunities for improvement
- Keeping an open mind
- Showing a backbone
- Confidence
- That real worthwhile change comes hard

In part 2 of this book, we will now examine *Project-Driven Change Leadership* and even more important, how best to achieve it. These important characteristics and much more will all now be explored and analyzed.

Part Two:

Achieving Project-Driven Change Leadership

A Brief Introduction

Leadership requires that someone lead; at the same time, others follow. It takes a certain kind of person to be a leader, meaning someone willing and able to do so. This job is demanding, and it is not intended for everyone. It requires very long hours, persistence, patience, a clear focus on the desired goal, and good and fair judgment. Charisma is useful, but not required. Excellent oratory skills are helpful, but not mandatory. Leadership sometimes comes from a person's genes, but usually, it is an acquired or learned skill.

Project leadership is specific to driving a team of skilled people through an endeavor with a specific purpose for some duration and producing a desirable result. Although it is closely associated with project management, it is somewhat different. Project management has more to do with managing a project's schedule, quality, budget, resources, and so on. Project leadership focuses much more on the human resource elements of project management. It has to do with matters such as effective conflict resolution, decision making, negotiating, and communicating. Projects are very stressful endeavors, and too often, emotional distress can outwit years of acquired knowledge, competence, and experience. Interpersonal or "soft" skills matter, and project leadership or the lack clearly demonstrates that by minimizing or ignoring any people elements in the management of projects can be costly and risky. Many project studies over the last several decades clearly indicate that people challenges are substantial and can make or even break an effort.

Change leadership focuses on driving people through an endeavor that challenges the status quo to achieve breakthrough best practice results. People tend to fear change, and too many far too often find countless ways to resist it. Despite this fact, if the change is for the better, it is well worth any effort required to bring it to fruition. Change leadership is about managing people from an As Is situation in which they now operate to the better To Be possibilities awaiting them. No one has ever said that this is an easy or quick process, but it is possible to achieve.

In short, managing yourself at times can get
interesting, leading others is usually challenging,
and driving people toward making a real and lasting
change is a significant hurdle to overcome, but
again, possible. By adhering to certain key principles
and following a clear and proven path, the odds for
achieving success in this challenging journey can
be greatly increased. First, align your actions with
your values. Then, prioritize and focus on what
is most crucial. Commit to continuous learning,
professionally challenging things, and personal
growth. Then, set and periodically review your goals.
Finally, frequently measure your results to assess
your progress or the lack. **Project-Driven Change
Leadership** is doable if one is committed to the
endeavor and acts accordingly. Last, recognize that
one cannot do it alone. It requires many followers as
committed to the cause and follow-through as need
be.

CHAPTER SEVEN

Align Actions with Values

"**A**ctions speak louder than words" is an expression we have heard from our parents, teachers, employers, and society throughout our lives. Although this statement is true, actions alone will not suffice. One must also "think before and while acting," another such common-sense expression. By so doing, knowingly or otherwise, our basic values come into play. Our value system is part of who we are as well as how we view the world around us. The experts who study values indicate that our values are locked by the time we reach ten, and they can only be altered by some life-changing event.

Now, in a business enterprise, we deal not only with the individual values of all employees, but even more with an overall organizational culture. Corporate culture has significant sway over employees and their managers. For instance, although business organizations have many policies and procedures, often, the unwritten ones have the greatest impact on employee and manager behaviors. No matter what, values are very important and key to obtaining actions that can then lead to the desired outcome.

"Values are judgments about what is most important in life and are an integral part of every culture. They tell people what is good, beneficial, important, useful, beautiful, desirable, and appropriate. They also help us answer the question of why people do what they do. Finally, values can be positive or negative, and even destructive."[1] They are very potent influences on human behavior. "Corporate culture is the total sum of the values, customs, traditions, and meanings that make a company unique."[2] Finally, *alignment* means "the adjustment of an object in relation with other objects."[3] The purpose of doing this is to optimize effectiveness, for instance, when we seek to maximize business value from using information technology, such as with an ERP system.

Success is more likely to come when we align our actions with our core values. For instance, in the first short story, the value of persistence paid off as the new division controller finally discovered the solution to the financial problem encountered after the new computerized drilling machine became operational. In the second short story, the value of willpower by doing what was best for the company overcame the CEO's resistance to the effort's high cost. In the third story, the consultant got the

[1] Wikipedia—Value (personal and cultural)

[2] Wikipedia—Organizational culture

[3] Wikipedia—Alignment

reluctant CEO to change his long-standing and strongly held position because he appealed to the CEO's basic human values about his concern for his new senior managers as well as the overall welfare of the business. On the other hand, by not aligning our actions with our values, we create unnecessary dissonance that hampers our best intentions and good efforts.

We must first understand our core values because we need to use them to set out on the right path to begin with so we can better achieve our objectives. By appealing to people's good inclinations to do the correct things, we tend to improve the chances of what we set out to achieve. While this idea makes sense, too many too often fail to heed the obvious, thus making things more difficult than need be, increasing the degree of risk, and placing unnecessary hurdles in the way that then need to be overcome. However, for those who pay close attention and follow along, the challenge faced is diminished, and the effort required for a successful result is diminished. Nothing guarantees success until the endeavor is completed, but by aligning the actions taken with our values, or most of those values, in a business enterprise, the likelihood of achieving the predetermined goals is more probable at much lower risk, cost, and effort. Actions taken in congruence with our values are just the first step down the path to a positive outcome. When actions are undertaken without regard for our values, extra effort usually results, and such actions will not ensure that we reach our desired destination. We must drive our actions based on a solid set of core values, for without this, any actions undertaken could be wasted. Given the pace of technological and other changes in the existing global business environment, this might mean the difference between survival and otherwise. Speed of action is a crucial success factor; however, doing the right things, and not what is expedient, matters greatly as well.

CHAPTER EIGHT

Prioritize and Focus

L eadership usually shines best when things are difficult, but it doesn't just happen when there is a significant challenge to be faced. It needs to be honed over time during the good and difficult periods. It begins with a vision of some desired destination to be reached in the near term. This vision or destination is some worthy business goal that justifies all the required effort and cost, including any opportunity costs. However, there usually exist many worthy objectives, so this presents us with a quandary. For some people, this analysis can become a paralyzing experience resulting in utter exhaustion, significant frustration, or possibly both, with no progress achieved. Therefore, first determining and then setting priorities is the best way to get things going and then to help keep them on track. A vision statement should represent an organization's priorities. The exercise of filtering the various worthy goals down to a common shared vision and destination is neither easy nor quick if it is done well. There is much to consider, not to mention the added task of obtaining all the proper buy-in and acceptance necessary to proceed properly to begin with. Organizational vision statements work best when they are clear, thus preventing any misunderstandings. The use of pictures, and not just words, enhances proper understanding by all for a much better achievement of the desired results.

Whether it is in a person's life or in the wider world of business, setting and then sticking to one's priorities is crucial to arrive at the desired future state. When we consciously consider all the alternatives and further take whatever time is necessary to do so, we increase the odds of getting what we want and when we want it. I am constantly amazed by how many people I regularly encounter who do not understand nor seem to care much, if at all, about priorities. Instead of setting a destination and then figuring out how to get there, they proceed right into doing certain things. What they do not seem to get is that their actions define their priorities, whether they realize it. This is also true even if we set priorities and then ignore them at the first available opportunity. By focusing on our priorities, we tend to stick to them, even more if we write them. But to focus also requires effort, sometimes much effort, especially when we are bombarded with many other interesting stimuli.

So, first by prioritizing and then by focusing, we improve the odds for achieving our desired goals. This is as true for people as for entire global business enterprises. In a business organization where many people with different values come together to work toward a common set of objectives, the challenge involved is substantially much greater. Despite this, by continually and consistently communicating what the goal is and why it is so important to achieve it, with making it a priority by focusing on what needs to happen and precisely when to reach the desired destination, the likelihood of a positive outcome is enhanced. It is too easy to get distracted or even worse, to give up on our goals. Staying focused over time is no easy task.

For instance, in the first short story, the priority focused on was fixing the financial problem resulting from the wild swings in the work-in-process inventory from period to period. In the fourth short story, despite the unexpected finding of the worker-to-supervisor ratio, by first focusing on the project's priority, the high maintenance of the existing standard cost accounting system, the consultant had the credibility to address this secondary, but still important, business matter. In the fifth short story, the bank forced the company's four executive brothers to give their full attention and energies to this company's pressing financial condition. Their sister added the extra spark that changed the bad behavior of her four brothers for the overall betterment of the business, its middle managers, and ultimately, all its employees and creditors, mainly the bank.

Focusing intently on the objectives sought is also important. When you combine this with the preceding effort to set clear priorities, you increase the likelihood for obtaining the desired result. Setting priorities requires effort. Holding to them is neither easy nor automatic. But if you can do so and then stay focused on the mission, you will more likely achieve what you set out to achieve.

CHAPTER NINE

Learn, Challenge, and Grow

L earning was never meant to be a onetime spectator event for just the early years of a person's lifetime. Continuous learning is required throughout one's life, and even more today, given the rapidly changing world of the twenty-first century. Specific knowledge and skills can quickly become obsolete, or at the least, somewhat diminished. The only good solution is to embrace continuous learning to survive and flourish in such an environment. Although formal education is necessary and worthwhile, it alone is inadequate. Learning comes from doing, not just from listening to some highly qualified teacher explain things to us in a classroom. However, doing things entails the risk of embarrassment with possible failure or achieving something somewhat less than successful. Yet formal learning is not nearly enough. It is only meant to be the beginning of our journey. When an entire business enterprise completely embraces continuous learning and makes it routine, they set a course for themselves that can only be described as beneficial.

Challenges come in many forms and on a timetable that is not our choice. In a project's life cycle, a project manager typically faces three types of challenges: people, process, and technology. In my previous book, *People Centric Project Management*, I shared with my readers some research spanning several decades that clearly indicated that people challenges exceeded the combined challenges of process and technology by a factor of four times. Nonetheless, all challenges need to be addressed and not ignored if we are to achieve success with the endeavor at hand.

Long-lasting growth arises from meeting the various challenges we face by applying our knowledge, experience, and full energies to the cause. Creating sustainable change from a project-driven effort is not a matter of good fortune; it takes much hard work and a bit of time. Rarely does a breakthrough result from our first or early efforts. It takes much patience with much focused persistence to hit the desired target. I have heard that for you to become proficient in a sport or any other worthy pursuit in life, you need to dedicate at least 10,000 hours to a combination of learning and practice. If the typical American worker spends forty hours a week for 52 weeks each year, amounting to 2,080 hours, this means that proficiency is achievable after spending at least five years on a sustained effort.

In the first short story, the new division controller had to invest some effort to get firsthand knowledge and experience that could then be focused on the significant financial challenge this division faced. In the second and third short stories, the substantial challenge came from the respective company CEOs. Although their individual objections were somewhat different, the result was the same, which was that they did not support the position taken by those around them, including their senior management staffs.

The ability to solve a problem requires more than just getting academic and basic knowledge. It encompasses the full application of creativity, a willingness to take a calculated risk or two, clarity of the objectives sought, a sense of timing (as in when to act or respond, as well as when to hold back from an inclination to do so), and finally on how to approach the challenge encountered. Many people obtain a formal education and get very good knowledge and skills; however, too few learn how to apply all this information effectively to deal with the challenges they often face, especially as regards achieving positive and sustainable organizational change. Finally, it is worth noting that "The Balanced Scorecard" approach first developed in 1990 as a strategic performance management tool by Dr. Robert A. Kaplan and David P. Norton contains four key measurements. The last of these is called "Learning and Growth." By doing, not just studying, things, we learn what works well and what does not. If we pay close attention by capturing the Lessons Learned from our project experiences, we prepare ourselves to do a better job in the future. Over time, this effort will then lead to our growth and development as we improve.

CHAPTER TEN

Set and Review Goals

The needs to determine, set, and periodically review goals are fundamental to every business enterprise. This fact is shown by every Business Management 101 college course regularly taught worldwide. Goals must be time boxed, and they need to be as specific as they can possibly be. A goal means "a projected state of affairs one intends to achieve."[1] Goals need to be measurable so we can subsequently compare our results to each goal and, if need be, make any mid course adjustments or corrections to get back on track toward our stated objectives. Further, goals can be tagged as either short term or long term. These periods also need to be clearly explained for any of those involved in setting and then working to achieve the intended objectives. Organizational goals are typically determined and set for each fiscal or business year and often for much longer periods such as three, five, ten, even twenty, years hence. Goals must be locked and taken seriously, but that

[1] Wikipedia—Goal

does not mean they cannot change if circumstances so warrant. However, changing a goal must be considered, well justified, tightly controlled, and then clearly communicated to all necessary parties.

Goals made in a business enterprise can be determined for specific business purposes, for instance, for certain business functions or departments such as sales, operations, finance, and human resources. Within these organizational elements of a company, the goals should focus on financial, operational, customer, supplier, and employee factors. Internal and external activities should have well-defined objectives attached to each of these factors. The premise behind proper goal determination and setting is that you cannot successfully arrive at the desired destination unless you first plan how to get there. It doesn't happen by accident or coincidence. It requires deliberate planning coupled with tightly controlled actions.

Objectives to be achieved for project management efforts should be obvious. Adhering to the project's schedule, spending within the project's approved budget, and meeting, or even better exceeding, in satisfying the end customer or recipient of the endeavor are the most typical goals. But what about important matters such as the things learned in the effort's performance that can be reused for a future project, the team building resulting from participating in the project, or the discipline acquired from selecting and then following a project management methodology throughout? Such experiences are invaluable in driving significant business change. The results achieved enable us to learn what works from what does not. Although setting performance goals is both good and useful, it is mostly about obtaining the desired outcome in the desired period. A result means "a consequence, score, or number."[2]

Project leadership first requires that a clear future vision be established. You can't lead people toward some nebulous place and expect them to arrive there successfully. Concrete written goals must exist so everyone involved first understands where you are driving to and why. Thereafter, all can clearly know whether the objectives were achieved. Periodically, it is fine to review your stated goals to make sure they are still valid; however, changing them requires predetermined, controlled, decision making. Finally, it is important that the project leader passionately believe in the set goals if he is to have any chance of getting others to help him achieve them.

[2] Bing Dictionary—Result

Leading Project-Driven Change

Leading Project-Driven Change

Leading Project-Driven Change

CHAPTER ELEVEN

Measure Results

It is imperative that you objectively and periodically measure results to determine whether progress is made toward the stated goals. The point of setting goals to begin with is to work in a manner that best enables you to achieve them. Without clear measurements, you cannot determine whether the goals are achieved. When we consider setting objectives, we should simultaneously determine the relevant performance measurements to use thereafter during the project's life cycle. The goals we set become targets to be achieved. Measuring our progress toward such goals helps us stay focused on the task, encourages continued good performance, and provides milestones at specific points to make sure we stay on course and celebrate our successes. Furthermore, by trending our results, we can more clearly record and review from where we have come, which becomes important years later when we seek to show exactly where we have come from to arrive at where we are and desire to get to. Finally, we can benchmark our results to other external parties doing the same or similar activities. This is powerful and useful information as well.

Your ability to measure performance objectively is a basic and required management responsibility. Determining what and when to measure are crucial success factors in project management. Although many good tools and methods exist for this purpose, knowing what to track and exactly how often are just as important. Keeping score alone contributes a discipline that helps the project manager keep everyone focused on what matters most. This is especially true for information technology endeavors where project failure rates remain stubbornly high. Performance measurement further provides information from which to determine any rewards given out and when.

Project leadership requires you to first set goals and then measure results. These steps are necessary to keep people working on the effort and those supporting it well informed throughout the endeavor. People need to hear from the project manager with as much clarity and objectivity as is possible how things are proceeding. The project manager wants to report good news as much as the audience wants to hear positive things about the effort. But if there is no basis in fact for such good news, then you are set up for future failure. Only when objective facts are used to express your progress toward your set goals is there good reason to celebrate.

Selecting the right measures to use in reporting results is not necessarily easy. Finding relevant measures sometimes can be challenging. In certain cases, it is clear-cut, as in measuring financial performance. Anything tangible makes this effort far less a burden than it would be otherwise. But intangible measures such as increased information for better management decision making can be difficult to nail down. In addition, if there is no performance baseline to compare current results to, then progress or the lack is not provable. Subjective thoughts are nice, but objective results are far better to use for basing any future decisions and actions. Measuring relevant results against your objectives on some predetermined time frequency is crucial in project management.

CHAPTER TWELVE

Project-Driven Change Leadership

The nature of work globally has shifted over the last few decades from an environment where people operate in independent functional or technical departments to one where people work in cross-functional and technical teams on project-driven activities to affect significant business change for the better. For instance, the idea of "best-practice" business processes speaks directly to this fact. Business enterprises worldwide strive to achieve this goal lately with and without the application of information technology, such as an ERP software system, to help in this effort. As such, to achieve breakthrough changes in their business processes, besides applying information technology devices, they are creating project teams to drive the goals they seek because they understand that cross-functional teams of people can potentially achieve far more and with much better results than otherwise. Project teams used to be formed to create

115

products and services. Although this is still the case, business enterprises apply the project and team approach to many other initiatives, including those involving improved information technology solutions.

The widespread activity of creating projects and teams to work them is now more of the norm instead of the exception, which is why the profession of project management is growing so significantly. Even more important, it is projected to create 1,200,000 jobs on average for each of the next six years, making it one of the fastest growing professions on the planet. Projects and teams are assembled to effect change. Aside from all the many tools, methods, and approaches available to a project manager, the most crucial success factor in significantly changing things has more to do with people. It takes leadership to break down any resistance to change typically encountered because too many people initially tend to fear change instead of welcoming it. So not only has this business trend increased the interest and demand for good project management practices and project managers, but it has also spawned a related profession known as change management.

Change management means "a structured approach to transitioning individuals, teams, and organizations from a current state to a desired future state."[1] Such change can be driven by some strategic objective, desired technological improvement, structural reorganization, or just changing employee attitudes and behaviors. Whichever change types it is, resistance will exist that needs to be effectively dealt with. The principle

[1] Wikipedia—Change Management

in change management is that if you understand the why behind the desired change, you can better achieve the how of getting it done successfully. If we understand and truly believe in the proposed change, we are then much better equipped to do all that is necessary to help bring it about. A side benefit is that we become a champion of the desired change sought. Then, if we couple this with project management, where some sort of change is involved, we have created a powerful combination if applied properly.

Today, projects are usually the means to effect significant business change. Both project and change management have methodologies and various tools behind them for project and change managers to apply as they see fit. Both involve some process to get from point A to Z by following a disciplined and controlled series of purposeful steps. Both need project and change managers to be leaders for best results. In many ways, they follow a similar path, first creating a sense of urgency to get the effort approved and then under way. Then, they need an executive sponsor and a team of professionals to help them move the endeavor forward. A clear vision of the future state must be developed and communicated consistently in the future. As the project moves along, various obstacles will be encountered that must all be effectively dealt with. Some of these impediments or risks were likely foreseen, whereas others might be a surprise. Given human nature and attention spans, the need to create short-term milestones where success can be recognized is crucial for ultimate success. Adhering to the best practice of collecting Lessons Learned for future use further shows the seriousness of the effort undertaken. Incorporating the project or change result into the organizational culture helps cement it eventually.

Another way of examining these two hot business ideas is to think of project management as primarily focusing on the technical elements of getting from a current to some better future state. By contrast, change management addresses the people and organizational elements of achieving the same objective. When these two concepts are applied simultaneously in a project-driven effort, the results often tend to be much better, which can be attributed to recognizing that people challenges are the greatest obstacle to success in a change effort. Methods, tools, and new technology present challenges, but if professionals work well together to solve the matter, they will get it done. On the other hand, if people challenges are raging, dealing with any other challenges becomes secondary at best. And if people challenges cannot be resolved, the endeavor is most likely headed to fail or end early.

When project and change management operate together, they still require project and change leadership. Without any leadership, the effort will experience a much more difficult time of it, not to mention having increased risk associated with it. In both cases, the global need for **Project-Driven Change Leadership** is great and expands. Project managers who recognize and embrace this key fact are not only enlightened, but also are more likely to incorporate both concepts into their daily endeavors to maximize the chances of a more successful and long-lasting outcome.

Part Three:

One Large Fable

CHAPTER THIRTEEN

In the Beginning

Once, a global business enterprise decided they needed an ERP system to drive a significant change in the way they now performed their business operations. Revenue was barely growing; their market share was declining in their key customer market segments; and their profitability was running flat despite several rounds of cost-cutting measures. They heard of a hot global ERP system taking the worldwide software marketplace by storm and already being used by some of their toughest long-standing competitors. So, they quickly put together a business case, allocated the required funds, and then obtained executive management approval to launch a project to deploy this ERP solution as the driver for both rapid and substantial business change.

The project was provided with a young hotshot project manager, Anthony, an up and coming IT middle manager. The project sponsor, his boss, advised him to evaluate and select an outside consulting firm to help him in this important endeavor. The project sponsor, Roger, the company's CIO, had recently come from one of the largest global IT consulting shops. Anthony asked Roger for some advice about how to go about evaluating and selecting a consulting firm partner for this project he was now about to manage. Roger told Anthony to "just go hire my former employer." When Anthony asked why, Roger shot back, "because they are simply the biggest player and the best, plus they have a strategic global alliance with the ERP vendor that their company had just chosen to use." So, Anthony decided that this big chore could be easily crossed off his to-do list.

He went right back to his office and located the website for this global consulting firm. He found an office in the city in which his company headquarters were, listing a partner's name, phone number, and e-mail address. Anthony then called George, the partner named in the website. Anthony briefly explained his need, and George offered to meet with Anthony in a few days. Anthony now felt confident about this important aspect of the project, so he turned his attention to other pressing matters. He saw Roger the next day to discuss the budget for this new effort. Roger said that Anthony should wait until George could provide his input. Then, Anthony asked Roger about staffing the project with company employees and managers, and again Roger deferred him to George.

At the meeting with George, Anthony laid out an agenda addressing a project team's staffing, a project budget, and a project schedule. George said he came prepared with solid and detailed answers to these and many other crucial project matters. George then explained to Anthony that he and his consulting firm had done this kind of endeavor countless times before, and they thoroughly understood all the details involved. All that Anthony had to do was trust George and the experienced consultants he was about to bring on this project, and all would be well. Anthony was a bit taken aback by how easy this key task had now become for him, but then, he supposed that was why these consultants are paid a great deal of money.

Within days, George introduced Anthony to Bernard who was new to this consulting firm and whom he said would be this consulting firm's project manager for this effort. He presented Anthony with two documents on an overhead projection slide show. The first of these documents was a project plan done using MS Project, displaying a list of tasks grouped into various phases representing all the project's planned activities, which impressed Anthony. The next document was a 183-page Statement of Work (SOW) that explained how the project would be conducted. Included in the SOW were many legal terms and conditions and sales and marketing information. It also included the high cost to this company for this consulting firm's services. It further named twelve consultants with their project roles and hourly billable rates. The SOW mentioned that each consultant would need to travel to this company's headquarters each week for the endeavor's one-and-a-half-year duration. Finally, it stated that this company would be charged for these services on a time and materials basis.

Anthony only viewed the documents; he was not provided electronic or printed copies of them then. George said they were still tweaking them before they would shortly be released in final form. George then told Anthony that Roger had already been briefed at a high level about the contents of these documents and that he had given his verbal approval to him. Thus, all Anthony now had to do was to sign these draft work-in-process versions in the appropriate places for the work to start. Anthony did just that right then, thinking that if Roger were comfortable, then so was he. So, he signed as required on the dotted line as George instructed him to do.

Next to be dealt with was the selection of all necessary company project-team members. Again, Roger came to Anthony's rescue. Roger told Anthony that for each consultant listed in the consulting firm's SOW, and twelve were mentioned, the company needed to match each consultant with five people. Before Anthony could ask why, Roger explained that he had used this rule of thumb when he had worked for this consulting firm. So, Anthony would need to rustle at least sixty company project-team members and do so quickly. Roger also strongly suggested to Anthony that at least half this number be IT staff, with the other half coming from the business side of the company. This, too, was another of those rules of thumb according to Roger.

At this point in their conversation, Anthony thanked Roger for all his good advice. Just as he was about to leave Roger's office, Roger told him that the company project team had already been determined for him. Roger said that he had just completed the entire decision-making process about the thirty members from the IT staff. Further, Roger then told

Anthony that he had just secured agreements with the vice presidents of sales, operations, finance, and human resources for the remaining business personnel for the project team. Anthony now thanked Roger for making such a daunting task so fast and easy for him. Roger replied, "It was my pleasure to lend you a helping hand and to play consultant for a bit again."

Anthony left Roger's office thinking that all this project management stuff was not so bad after all. His little research on the Internet had indicated to him that project management was a significant responsibility and challenge; however, it was not proving true in this situation. Given the speed and ease with which he had achieved so much to date, especially in getting some first crucial pieces put in place, he thought the rest of the project would go likewise.

Taking Bernard's advice, Anthony called the sixty company staff and twelve consultants together for a project kickoff event. Bernard prepared the slide deck that communicated the project time line, milestones, deliverables, and budget. A catered breakfast was provided to make everyone feel as comfortable as possible. Bernard predetermined that Anthony would introduce this ninety-minute presentation, Bernard would go through most of the 170 slides, and Anthony would close the event with a brief question-and-answer period.

The kickoff event was held first thing on a Monday morning. Roger, the project sponsor, had a last minute, out-of-town business meeting to attend, and thus, he was unavailable to participate and offer his words of wisdom to the assembled group. The

vice president of sales was away with his wife on
a trip he had earned for last quarter's better than
expected sales performance. The vice president of
operations had planned to stop by, but an emergency
in the plant prevented him from doing so. The vice
president of finance was immersed in quarter end
closing activities, whereas the vice president of
human resources was attending a human resource
professional conference out of the country. Although
Anthony would have preferred that all or at least
some of them be present for this event, he realized
that herding executives was beyond his control.
Anyway, he knew that he was in able hands thanks
to Bernard.

The kickoff meeting seemed to make everyone
present feel good, and it was completed fifteen
minutes earlier than it was called for, which pleased
them all as well. This, even though it took twenty-
five minutes just to have all seventy-two people and
the two project managers introduce themselves at
the start of the meeting. Further, hardly anyone
had any questions to ask Bernard or Anthony. As
the group broke up, Anthony was feeling terrific
about getting the endeavor, which was to begin that
afternoon, under way.

The first project working meeting was with all
thirty of the company's IT project team members and
all twelve consultants. However, only eleven of the
thirty company's IT project team members showed
up for the meeting. This poor showing surprised
Anthony, but he knew many of his fellow IT staffers
were busy helping with the financial quarter end
closing, particularly given the way their various
information systems and business processes now

operated. He chalked this up as a onetime thing, and he believed the eleven who showed up would update the other nineteen people soon.

Leading Project-Driven Change

CHAPTER FOURTEEN

In the Middle

The next day, Anthony came in to work a bit earlier than usual because he wanted to ask Bernard about something important. Anthony had seen an advertisement in one of his information technology periodicals about a global professional organization called the Project Management Institute (PMI). He wanted to get Bernard's opinion about this professional organization before he approached Roger to explore the possibility of becoming a PMI member. When he caught up with Bernard and asked him about PMI, Bernard quickly responded by saying that "he never heard of this professional organization." With that said, Anthony immediately put this idea out of his mind so he could better focus on the more crucial matters having to do with his company's project for the balance of the day.

According to Bernard's project plan, the next significant task had to do with gathering information about the company's current environment of business processes and information management systems. Bernard's right-hand assistant for starting this important activity was Alice. She now asked Anthony to call the entire project team together so she could explain to them how this task was to be handled. Anthony did as she requested, and about one third of the company project team showed up for Alice's PowerPoint presentation. All consultants were on hand. Anthony was embarrassed and disappointed by this poor turnout. But to forge ahead with the project, he asked Alice to redo her presentation and to e-mail it to all missing company project-team members. Alice protested Anthony's request, but Bernard interceded in support of Anthony, so she did as she was instructed.

Now that Alice's presentation was completed, she abruptly left the meeting room and was not seen again for the rest of the day. Anthony took off as well, but his mission was to learn why so few project team members showed up for Alice's presentation. His first stop was to see his manager, Roger, about his fellow information technology staff members. Roger told him that as far as the existing environment was concerned, this activity was primarily meant for the business people, which is why he had instructed the information-technology project-team members not to participate. Anyway, he further added, the information technology landscape was documented in detail because of a recent requirement made by the external financial auditors. So, Roger told Anthony that the Information Technology Department was well prepared in this regard and not to worry about it.

So, with Roger's assurances about the Information Technology Department, Anthony struck out to see why so many other project team members from the business side failed to attend Alice's important presentation. He first encountered the vice president of sales in the office hallway. Anthony asked him why so few of his staff assigned to the project were present at Alice's presentation on how the upcoming As Is task would be conducted. The vice president of sales said that he was unaware of such a meeting, probably because of his heavy travel schedule, but he would have his administrative assistant look into it and then get back to Anthony.

Anthony next encountered the vice president of operations and again raised the same question with him. The vice president of operations told Anthony that there had been many plant issues lately to deal with, and these could continue for a long while so Anthony had better realize this fact of life.

Finally, the vice president of finance reminded Anthony about the year-end closing still under way. At this point, Anthony gave up and headed back to see Roger. He was feeling frustrated about the apparent lack of interest in, and support for, the project. Roger told him "not to worry; it's just the very early part of the endeavor and this kind of thing is to be expected, plus anyway, most of the departments had accurate and current documentation of their existing business processes thanks to the outside auditors." Roger further explained to Anthony, "This portion of the effort was primarily for the benefit of the consultants in order for them to get up to speed on the company's existing business practices." Roger finally told Anthony just "to chill, and all would be fine." So, with his manager's words of wisdom, Anthony began to feel much more comfortable about the project again.

The next two weeks proceeded the same way. Varying numbers of company project-team members were involved in the sessions held to discuss the existing business environment. Sometimes, it was a bit more, whereas on other occasions, it was a bit less, averaging about 30%. Bernard mentioned this matter to Anthony a few times during this time, but Anthony indicated that once the project shifted to the next major activity, which was developing the future state business and information-management system environment, attendance would improve.

The As Is sessions conducted went fairly well. The current business and information management systems landscape was explained in detail using Visio diagrams supported by many other relevant materials. At the end of these two weeks, Bernard approached Anthony to tell him that Alice had just resigned from their consulting firm, and she was headed for greener pastures. Thus, she would no longer be available to deliver the upcoming presentation having to do with jump-starting the To Be effort. She had her slide deck already prepared for this event, so Bernard said he would deliver this presentation in her absence. Under these circumstances, Anthony felt relieved that Bernard was willing and able to step in so no further time delay would be incurred.

The meeting was announced with more than enough notice, and all the vice presidents and their administrative assistants were copied to remind them of this important event. Despite this fact, just 15% of the company project team bothered to show. Anthony could not believe this awful turnout. Bernard asked Anthony if he should postpone the session about to start.

Anthony excused himself to find Roger. He raced upstairs and to Roger's office and found the atmosphere strange. Roger was busy packing his personal items in a few company boxes. Anthony asked Roger "if everything was all right?" Roger answered him that he had just come from the CEO's office where he had announced his departure from the company. After learning of Roger's stated intention to jump ship, the CEO requested that Roger leave the company immediately, if not sooner, instead of waiting for the customary two-week notice to run its course.

Anthony was stunned, and he did not know what to say to Roger. His mind raced with many thoughts about Roger's news, its impact on the Information Technology Department, and its effect on the project now well under way. It took several minutes to sink in as Anthony suddenly remembered why he had come to see Roger. But before Anthony could utter a word, Roger realized that there was no possible way for Anthony to know about his sudden departure beforehand, so he asked Anthony why he had come to see him.

By this point, Anthony had regained some of his composure to share with Roger the news about the terrible attendance at the meeting downstairs in the large conference room. Roger politely listened and then told Anthony, "Well, I just can't help you anymore, but do your best and trust Bernard, and all will work out well." Roger then asked Anthony to leave him so he could finish packing his things and then get to human resources to end his employment at the company. Anthony was again stunned, but he followed Roger's wishes as he remembered about the project meeting downstairs still waiting for him to return.

Anthony finally returned to the suspended meeting to face Bernard. Bernard was annoyed but now suddenly sensed that something was wrong by looking at Anthony's facial expression. Despite this, Bernard then said to Anthony that the small assembled group was still waiting for Anthony to determine whether the To Be jump-start event was still on. Anthony blurted, "Roger is leaving the company as I speak."

Bernard and all the others in the large conference room within earshot became visibly upset by this news. "This can't be," Bernard responded, "he is the project sponsor, yet alone the company's CIO." Anthony said, "I am afraid it is true. Roger just announced his departure to the CEO, who then dismissed him right on the spot." Bernard regained some of his composure to ask, "Where is Roger going?" Anthony replied that in all the confusion, he forgot to ask. Bernard finally suggested that, given the low turnout for this event and the news about Roger, the meeting be postponed to another day. Anthony just shrugged his acceptance of this idea, and Bernard told all present of this decision. Anthony barely heard a word Bernard uttered.

Anthony came to work the next day still feeling shell-shocked. He opened his computer to check his Outlook calendar and e-mail and spotted an e-mail from Roger. Roger told Anthony in his e-mail that he was returning to his former employer, the consulting firm now helping this company in this crucial endeavor. Anthony was dumbfounded.

Shortly thereafter, he saw Bernard and asked him about this matter. Bernard told Anthony that he learned about this piece of news late last night from George. George told Bernard that Roger was returning to their consulting firm because he missed the thrill of being a consultant as well as managing other consultants. Anthony then asked Bernard if he knew whether Roger would be involved at all in their project in the future, and Bernard replied probably not, given the position of the company's CEO toward Roger.

Anthony then decided to put Roger behind him so he could get back to the business of this project. He turned his attention to the postponed To Be kickoff event by asking Bernard for his advice. Bernard recommended that Anthony revisit the various company vice presidents to remind them of how important this meeting was and what was to follow and to get their commitments to push their staff representing the business side of the company project team to show up and take it seriously. Anthony agreed to do so right away.

Anthony sought the vice president of sales. He found him out of town on a business trip, so he communicated all that Bernard had just conveyed to him to the vice president's administrative assistant. She took down the information and then promised to deliver it later that day when he called to check things.

Next, Anthony went to see the vice president of operations. His administrative assistant told him that he was in the plant taking care of another emergency, but she took down the information and promised to deliver it to him as soon as possible. Anthony went to see the vice president of finance who was in a meeting with the external auditors. So, he again delivered his message to the vice president's administrative assistant. Finally, he went to see the vice president of human resources who was also out of the office attending a seminar. Again, the vice president's administrative assistant took down the message and promised to communicate it sometime later that day.

Anthony knew the administrative assistants would pass along the information, but he did not feel comfortable about it getting the proper reception and timeliness he was seeking, given the nature of the message. He returned to his office to compose an e-mail restating exactly what he had just communicated to all the vice president's administrative assistants, and he copied the CEO. In a separate e-mail, he asked the CEO with whom he should now communicate in the Information Technology Department given Roger's sudden absence. Now, all Anthony could do was wait for everyone's e-mail responses.

Meanwhile, another consultant announced that he, too, was leaving this consulting firm. Bernard reported the news to Anthony, and then, he indicated that a suitable replacement was being sought in their firm. Anthony had other more important and pressing things going through his mind about the project, so he quickly dispensed with this additional piece of bad news. He had confidence in Bernard and George to do the right thing for both the project and the company, their client.

The day passed with no responses to his two e-mails or to his visits with the various administrative assistants. The next day came and went with still no response to any of Anthony's actions. Bernard grew impatient, and Anthony now became concerned.

He again asked Bernard for his advice. Bernard told Anthony that, in the interest of time and to stay on plan, it was best to reschedule the To Be kickoff event two days hence. Anthony agreed, thinking that the results of his recent efforts would likely pan out by then anyway. So, the meeting was rescheduled as Bernard recommended.

These two days came and went, yet there was no response to any of Anthony's efforts. Despite this, the future state jump-start event was held. This time, 25% of the company project team was on hand. Just before Bernard began to deliver the slide deck prepared by Alice, Anthony interjected with a question addressed to just the company project team members present before him. He asked them if any of them knew why the rest of the project team was missing. Two people slowly spoke to say that many thought the project had been cancelled because Roger was no longer employed by their company. This answer surprised Anthony, and he was unsure how best to reply. Bernard jumped in to say, "Exactly how did this notion come about?" One person who had answered Anthony replied, "They just assumed this was the case."

Following Bernard's presentation, Anthony decided to take some personal project responsibility. He sent another e-mail, stating to all project team members and to all vice presidents and the CEO that until he was instructed otherwise, the project was still on and moving into the future state phase. Anthony further mentioned that all designated project team members were to participate regularly and fully in all project activities and meetings as required according to the project plan and the more detailed Excel schedule.

As a direct result of this action, at the first To Be brainstorming session on the sales order process the next day, 85% of the required project team members were on hand. Anthony's newfound take-charge attitude impressed Bernard. The results clearly spoke for themselves, as this meeting turned out to be a great success in generating a significant number of improvement ideas taking the features and functions of the new ERP system into account.

The next day, Bernard approached Anthony with some news of his own. Bernard said that he just found two replacement consultants in their firm to substitute for Alice and the other consultant who had suddenly left their firm, who affected this project. With all else going on, Anthony had almost forgotten this important matter and was now pleased to learn that these two replacement consultants would be on board within three days. He never thought to ask Bernard anything further about this matter.

Three days later, as Bernard had promised, Joshua and Cynthia showed up. Bernard introduced them to Anthony by saying that they were longtime consultants with their firm and that they came highly recommended, according to George. Anthony thought this was fine because George seemed to understand completely the nature of this project and company. Bernard indicated that Joshua would be ready to facilitate the first financials future-state brainstorming session that afternoon. Anthony was pleased to learn this. He had grown concerned about this session, so he was relieved to know that it was in able hands.

Anthony was needed elsewhere just as Joshua was about to begin the financials To Be session. Within ten minutes, it became apparent to many financial project team members present that Joshua knew nothing about accounting best-practice business processes. The company controller and several other financial managers demanded that Joshua take a time-out so they could find Anthony and convey this new situation to him.

The company controller interrupted another meeting to ask Anthony to step out to speak with him. He then explained to Anthony how awful Joshua was. This revelation shocked Anthony, and he raced off to find Bernard. Bernard was talking to Cynthia about the operations future-state brainstorming session for which she was busy preparing. Anthony asked Bernard to break away from this conversation to speak with him.

Bernard seemed annoyed, but he followed Anthony's request. Anthony was agitated as he told what the company controller had just told him about Joshua. As Bernard listened, he, too, became upset. Anthony wrapped up this brief discussion by asking Bernard what he planned to do about this situation, and when. Bernard replied that he would immediately find George. Anthony then said that he would instruct the company controller to return to this session led by Joshua to tell everyone that the meeting was postponed until further notice.

The next morning, Anthony sought Bernard to obtain an update about this matter. Bernard was nowhere to be found. Anthony waited another 45 minutes and then did the same thing. This time, he found Bernard with Cynthia again. Anthony interrupted them by asking Bernard what was happening about Joshua. Bernard shrugged and said that he was still waiting to hear back from George. Anthony then asked Bernard if he knew when George would do so, and Bernard said he did not know. Anthony was unhappy with this response, but he decided to give Bernard and George another 24 hours to get back to him. Before leaving Bernard, Anthony asked if Cynthia was ready to facilitate her first future-state session with the operations project team, scheduled for later that morning. Bernard said that she was well prepared, and Anthony then left them alone to continue their conversation.

A short time later, Cynthia got her operations To Be session under way. The vice president of operations decided at the last minute to show his presence, at least for the first fifteen minutes of this planned 2½-hour meeting. Anthony again had a pressing matter and could not attend the first 1½ hours of this session.

After twelve minutes, the vice president of operations asked Cynthia to stop this meeting so he could see Anthony about what was happening. He stormed out of the conference room, almost knocking down someone passing in the hallway as he raced to find Anthony. Despite his efforts, Anthony was nowhere to be found, so the vice president of operations returned to the To Be session to announce that he was cancelling it until further notice. He then returned to his office to instruct his administrative assistant to find Anthony as soon as possible and to summon him to the vice president of operations' office immediately.

A short time later, Anthony arrived at the vice president of operations' office. The vice president unloaded a tirade on Anthony about Cynthia's sheer incompetence. Anthony allowed the vice president to vent for several minutes before excusing himself.

Anthony demanded that Bernard come to his office right away. Once Bernard arrived, he told Bernard to call George's cell-phone number. Anthony's sudden request startled Bernard, but he did as Anthony demanded. George was at the airport waiting to board his flight when Bernard's call came. He took the call knowing that he had a very few minutes to speak to Bernard. When George answered, Anthony insisted that Bernard turn on his speakerphone so all three of them could communicate.

George spoke first and quickly explained his time constraint. Anthony said he would be brief and to the point, and then he told them that he wanted Joshua and Cynthia removed from their project immediately. When George asked him why, Anthony said because they have been deemed unqualified for the activities they were assigned. George then asked Bernard to gather more details from Anthony to share with him three hours later after his flight landed. After George hung up, Anthony calmed a notch to explain to Bernard all that had happened about Cynthia's now cancelled session.

The next day, George called Anthony to apologize. He explained that both Joshua and Cynthia had come highly recommended from another office of their consulting firm, but because of an administrative error, their résumés had been confused with two other consultants with the same first names. Once Anthony heard George use the word *résumés,* he asked George to send him the résumés for all the consultants now staffed on this project, including Bernard's résumé. George said he would have his administrative assistant e-mail them to Anthony within the next two hours.

After lunch, while checking his e-mail, Anthony spotted one from George. He opened it to discover the résumés of all the consultants now on this project. He selected Bernard's résumé to review first. As he read it, he quickly realized that although Bernard had some good ERP knowledge and experience, none of it was with the particular vendor solution to be implemented at this company. Further, he learned much to his surprise that Bernard was new to this consulting firm. This all disturbed Anthony, primarily because neither Bernard nor George had ever mentioned these facts to him. Then, as Anthony read all the other consultants' résumés, he found several others in the same circumstances.

He immediately called George to express his concerns about this matter, but George was unavailable to take his call. Anthony decided to let it go for now, but he told his newly discovered knowledge to Bernard. After listening to Anthony, Bernard said that George and Roger thought that if you had good ERP knowledge and experience, then it did not matter which vendor solution was deployed. Anthony again decided to let this matter pass for now because there were more pressing project activities to attend to right then, such as the future-state brainstorming sessions.

Bernard convinced Anthony that he could get the financial To Be session restarted while a suitable replacement for Joshua was found by his consulting firm. Anthony agreed to this, so this session occurred. At the brainstorming meeting, the corporate controller got into a serious dispute with a plant controller over how best to allocate certain overhead costs to the company's various products. Although each of them seemed to argue well for his different position on this matter, the tone of their disagreement was troubling.

Anthony heard about it from Bernard and decided to come to this session the next day for the continuation of the discussion. Anthony got a firsthand look at their conflict. This plant controller had been with the company his entire 28-year career whereas the corporate controller was new. Further, age differences only tended to aggravate this situation. They put forth strong cases for their different viewpoints as Anthony saw it. Today, their conflict grew in intensity as other financial project team members gravitated to one side or the other. It was fast becoming a mess, and Bernard was uncertain what to do about it. Anthony also seemed surprised that the two combatants could not work out their differences. Because of this, the financial brainstorming session abruptly halted.

Later that day, Anthony learned of another conflict, this time among the consultants. It concerned a template some of them wanted to use to capture the results of these To Be sessions, while others strongly objected to it in favor of another device for this purpose. The two warring factions split between those with knowledge and experience of the ERP system to be implemented at this company and those with a background in another vendor's ERP solution. Bernard was also torn between these two tools. This situation with the consultants exasperated Anthony. He returned to the office at day's end for one final check of his e-mail. Anthony spotted a new message from the company's CEO with a subject line stating "Very important." He opened this e-mail to learn that his company was soon merging with a longtime rival. Anthony was shocked by this unexpected news, and he began to wonder what it might mean for the project.

Chapter Fifteen

In the End

A few days later, as these two conflicts raged among the financial project team members and the consultants, Anthony received an e-mail from an old college friend reaching out to him through the Let's Stay in Touch website. He read the message rapidly and noticed that his old friend, Larry, was a certified Project Management Professional (PMP) and a self-employed project management consultant in town. Anthony found this interesting, but he moved on to more immediate issues concerning his project. A few hours later, as he sat through another meeting Bernard had called to try to resolve the dispute among the consultants, it dawned on Anthony that perhaps Larry could help. After this meeting broke up with still no resolution achieved, Anthony returned to his office to review Larry's e-mail again, which he had left in his Inbox. Larry had included his cell-phone number, so Anthony decided to call him right then.

Larry was very happy to hear from Anthony after so many years had passed since they were together in college. They spent some time catching up with each other's life journey because they had left college 7½ years earlier. Anthony then asked Larry about his project management consulting practice, and he explained his current role as a new project manager at his company. Larry suggested that they meet for breakfast the following week when he was back in town to continue their conversation. Anthony promptly agreed, and they set a time and place to do that.

At the appointed day and time, they met as planned. Larry asked Anthony to tell him more about his project situation. Anthony did so as Larry listened intently. Occasionally, Larry asked to clarify a particular point. Anthony felt more comfortable opening up to Larry as their time together passed. They spent ninety minutes before Anthony indicated that he needed to return to work. Larry then proposed an idea for Anthony to consider before they parted company. Larry asked Anthony to allow him to spend two days observing various daily project activities free to his company. Anthony said he would consider Larry's kind offer and get back to him within thirty-six hours.

Anthony arrived back at his office and called George to discuss Larry's proposal. He found George not particularly interested in this notion, but George told Anthony he would not object to it if that was what Anthony and his company's CEO decided to do. Anthony then visited the CEO to get his thoughts about Larry's offer. After listening to Anthony, the CEO encouraged him to move forward and get Larry involved. Anthony immediately contacted Larry to inform him, and they decided to have Larry come in later that week for two days to shadow Anthony as he went about his usual project management duties. Anthony then told this news to Bernard who, like George, was reluctant, but Bernard voiced no opposition to it.

Larry arrived as was planned and told Anthony that he would be as invisible as possible all day to observe everything that transpired for the project. Larry also told Anthony to tell people that Larry was a graduate student interested in learning firsthand about real-life project management. With that said, Anthony proceeded to the first appointment of the day. Bernard had called all the consultants together for another attempt to resolve their outstanding dispute about which tool to use to capture the results of the future-state brainstorming sessions. Factions were to present why their chosen device was best. George had instructed Bernard to close this matter one way or another as soon as possible. The combatant presenters rose to make their cases, explaining in detail why their selected tool was better than those of the other parties. Following the two presentations, Bernard indicated that he and Anthony would consider their arguments to decide finally by the close of business the next day.

The next event on Anthony's calendar for the day was to attend the financial teams' To Be session where the two warring managers continued their conflict. Each of these two financial managers was told to present his respective case in detail so Bernard and Anthony could make a final determination for the project to forge ahead. They all put forth their arguments, and the meeting was shortly adjourned. Anthony's final activity before lunch was to introduce Larry to the company CEO who had requested this when Anthony first mentioned Larry to him. The CEO asked Larry about his background and asked that Larry document his findings and recommendations from his two-day visit and then provide this information to himself and Anthony. Larry agreed without hesitation. The only thing he asked for was an opportunity to meet with the CEO and Anthony to go over his findings and recommendations in person. They then set a time for early the following week for Larry to come in to do that.

Anthony continued through his workday and the next one, attending to many other important project matters. At the close of the second day, Larry thanked Anthony for allowing him to observe everything, and he then headed off to develop his presentation of findings and recommendations. Meanwhile, Anthony struggled with the consultants fighting over which template to use while the two financial managers kept up their battle as well. Bernard tried to help Anthony with both matters, but he couldn't resolve either situation either way.

Now, it was time for Larry to return and present his findings and recommendations. Anthony did not know what to expect as Larry proceeded with his presentation to the company CEO and himself. Larry first indicated that from his two full days of observing the project by shadowing Anthony that he knew that this was Anthony's first project leadership experience. He went on to say that Anthony was trying his best, but he needed some help, especially given Roger's recent departure from the company. He then told the CEO that Anthony's knowledge and experience of the company were important to the project; however, being an effective project manager required much more than that.

Larry said that the consultants were leading Anthony and the project team members instead of Anthony driving them. He reminded them that an ERP implementation was primarily for the benefit of the business, not for the consulting firm, nor just the Information Technology Department. Larry took no stated position about the chosen consulting firm other than to say that the firm's reputation overall was good, but possibly, the particular consultants involved were not the best fit for this endeavor.

Larry then asked the CEO to allow him two additional weeks to delve deeper into the project and then return to elaborate on his findings and recommendations. The CEO again did not hesitate in granting this request, even though, this time around, Larry asked to be paid for his time and efforts. His fee was far more reasonable than what the consulting firm charged, so the CEO felt he had little to lose and possibly much more to gain from agreeing to this modest proposal.

After getting down to work, Larry requested many details from Bernard concerning this project. He interviewed all the consultants, including Bernard, after obtaining all their résumés from Anthony. Larry then met with key project team members, who for the most part, were middle managers. Finally, he spent some time with Anthony. Then, he again went off to develop his new findings and recommendations slide deck and shortly returned to present it.

Larry first wanted to share and explain it all to Anthony. Anthony was glad to have a first crack at hearing what advice Larry had to offer. Larry told him that the project started poorly. When Anthony asked for specifics, Larry was prepared. Larry told him that although the consulting firm selected was fine for the project involved, most of the individual consultants involved, including Bernard, were not. Larry explained that the particular ERP solution the company had chosen was far different and more complex from most other ERP packages, and these consultants, for the most part, lacked particular knowledge and experience with this specific ERP software. Furthermore, too many of them had never facilitated future-state brainstorming sessions, which was far more challenging than collecting information about the state of affairs.

Larry next addressed the project plan George and Bernard had originally put forth. He told Anthony that it lacked enough details throughout, and the planned time-lines were not realistic. As far as the project budget, based on actual spending to date, if this trend was projected forward at the same rate of spending, the result would be that this endeavor would cost four and a half times as much as what the original approved budget allowed. Finally, Larry concluded by saying that the project lacked quality standards, especially in key deliverables. Anthony took it all in for few minutes, and then he asked Larry if this project was destined to fail. Larry said that unless things changed, the likelihood of failure was high. Anthony then wanted to know whether anything could be done at this point that could correct this situation. Larry told him that he had developed a set of recommendations that, if applied quickly, would get things back on track, but the effort would still cost somewhat more and take a bit longer than their consulting firm originally stated.

A short time later, after Larry had a chance to share these findings and recommendations with the company's CEO, with Anthony present, it was decided to make Larry Anthony's right-hand assistant in the future. The CEO then told George at the consulting firm that Anthony would remain as the project manager, but Larry would be his chief advisor. The consulting firm would take their marching orders from Anthony, helped by Larry, from now on. George had no choice but to comply with this directive.

Within days, Bernard and several other consultants were removed from the project team. Anthony, based on Larry's advice, sought suitable replacements with certain skills that those asked to leave lacked. Larry reworked the project plan and budget with Anthony's full involvement, explaining his reason for every addition or change made to it. Anthony began to feel as if he understood what would happen, why, and when. Larry shared with Anthony additional project management information about the triangle that balances cost with time and quality around a well-defined project scope. It all made good sense to Anthony for this project. Larry also advised Anthony to seek more information by joining and participating in the Project Management Institute (PMI).

Within seven working days, the remaining consultants and project team members were assembled so Anthony could update them about the

Leading Project-Driven Change

many project changes in process or about to occur.
Many people asked many questions, and Anthony
could answer them all easily, effectively, and clearly
while Larry proudly looked on. Five working days
later, the consulting firm had suitable replacements
on-site at the company who were all now well
prepared to get to work immediately.

Anthony then relaunched the project, introduced
the project team members from the consulting
firm, and unveiled the revised project plan and
budget. The tasks already performed with suitable
quality, such as most As Is data collection efforts,
were noted as completed,. A brief period was now
provided to finish this work by those who needed
to do so. Then, the To Be workshop schedule and
approach were explained to everyone involved in it.
These brainstorming sessions proceeded as intended
with excellent results and near 100% participation
consistently.

As this project stage was wrapped up, the
company CEO announced that the merger with
their archrival was now being called off because
of insurmountable differences. Anthony viewed
this as good news for the project in that they could
concentrate on the business they already knew
so well instead of dealing with so many other
unknowns about their rival company. In addition,
project team members now would not need to be
diverted from their important project responsibilities
to handle any required extra activities associated
with a merger.

The project continued with a much clearer
understanding of what was required, a high level of
participation, and significant interest by all. Soon, it

became time to cut over from the legacy systems to the new ERP solution. "Go-live" came and went with few problems. The project now headed to its closeout. All the project documentation was placed in a shared network drive for future company reference. Lessons learned were gathered and discussed by all for potential future use as well.

Larry came to Anthony to say goodbye. His mission was achieved, and it now was time for him to move on. Anthony thanked him profusely for all his good advice and support. He told Larry that the total effort took 16% longer and cost 21% more than the original plan put together by George and Bernard, but under the circumstances they endured along the way, the company CEO was pleased with the overall results achieved. Before he left, the company CEO asked Larry for one last favor. He persuaded Larry to return in one year to conduct a brief post audit of this project. Larry immediately agreed to do so.

Exactly one year later, Larry returned as promised to this company for the express purpose of performing a project post audit as the CEO had requested. Before starting his work, he met with Anthony to catch up on things. Anthony proudly told Larry that since they were last together a year ago, he had become an active member of his local PMI chapter, had taken a Project Management Professional examination preparation course, and had passed this test on his first attempt. Larry was overjoyed for Anthony. They chatted a bit further about how things were going at the company with the new ERP system in operation. Anthony took Larry around to the various company vice presidents to get their takes on this matter. Each expressed great pleasure about the results achieved to date. For the most part, these outcomes were much greater than had originally been expected. They thanked Larry for helping them get back on track and seeing the endeavor through to its conclusion. Larry and Anthony couldn't have been happier about how things turned out. Larry wrote up his report, submitted it to Anthony and the company CEO, and then bid goodbye again to Anthony.

A short while later, the CEO created a Project Management Office (PMO) in the company and offered the position of manager to Anthony. He readily accepted and soon notified Larry of this latest turn of events. Larry wished him well with his new mission and told Anthony not to hesitate to contact him for any reason. Anthony said that he would stay in touch with Larry and again offered his sincere thanks for all Larry had done for him and for this company. His last comment to Larry was that he would look for him at the next lunch meeting of their local PMI chapter in two weeks.

CHAPTER SIXTEEN

Lessons Learned

This worst-case, large-project fable ended well considering all that transpired. In chapter 13, In the Beginning, there were many early mistakes made and far too much unearned trust given to the outside consultants.

- We were introduced to a project manager named Anthony, who had no project management knowledge and background. Like many others before him and most likely to follow him, he was given this responsibility because of circumstances. Given that his manager, Roger, was much more experienced for this endeavor, which clearly involved the deployment of new information technology, this did not seem so bad at the time. Anthony was a hardworking, dedicated, and loyal employee with much ambition and potential. The consulting firm employed also had very good credentials for this effort, which given Roger's past association with them, seemed a good omen as well. Despite all these positive

155

early signs, things turned sour and did so quickly.

- Roger's prior employment with the chosen consulting firm represented a potential conflict of interest. It caused Anthony to accept things more readily from them at face value and to trust that all would be fine, especially given Roger's supportive comments about them.

- George, although a consulting firm partner, was a consulting services salesperson with no specific background related to this project's requirements.

- Bernard had limited ERP background, just not with the software vendor chosen by this company.

- Anthony did not approve the other consultants brought on this endeavor. Their résumés were never requested and, thus, were not reviewed before their arrival.

- So because of these actions, whether intentional, the company rapidly became completely reliant on this consulting firm because no independent due diligence had been performed.

- Next, without prior opportunity to review the project plan and statement of work, Anthony concluded that he had to sign off on these documents. To make matters even worse, the project plan was still a work-in-process device.

- Like the lack of information about the backgrounds of the consultants brought on the project team, Anthony's failure to screen the company members was equally bad, if for no other reason than to know them beforehand. The other problem with these project team members was that they were not assigned to the project at 100% effort. Instead, most of them treated the project as an add-on activity.

- The project kickoff event and the materials presented contained no company member participation or input, sending a clear message that the outside consultants were driving the project.

- At this key project event, no company executive management presence was evident, further sending a clear signal to all other project team members that this endeavor was not that important despite what certain slides portrayed at this kickoff and what the consultant presenter said. Worse, the project sponsor, Roger, was nowhere to be found. That trend unfortunately continued.

- Anthony, the company's project manager, played a minor role at the kickoff event instead of leading this important first project gathering.

- Roger, the project sponsor, was from the information technology organization and not from the company's business side. Although technology was involved, business requirements and conditions primarily drove.

- The various rules of thumb Roger put forth and used to justify specific decisions made and actions taken, proved neither useful nor beneficial. These refer to the project team makeup between company and consulting firm members and company business and information technology members.

- Finally, the rapid-fire way the kickoff slide deck was presented was not conducive for raising questions and obtaining answers.

In the Middle, chapter 14, many more mistakes were made.

- The lack of participation by company project-team members not only embarrassed Anthony, but it also gave the project an early widespread poor reputation. For some time, it seemed that only a few company project-team members wanted to associate with this effort.

- The unannounced time for vacation and for company project-team members to attend outside conferences only added to this perception.

- No visible executive management support or presence shown further contributed to the idea that they weren't serious about this project.

- Then, Roger's sudden departure from this company to return to the consulting firm used for this important effort left Anthony completely exposed. Given his overall lack of knowledge and experience with such an endeavor, he was now in the hands of Bernard and George. Anthony would soon realize they were no better equipped than he was for this project.

- The next shoe to drop was the sudden departure of two consultants. If this was not bad enough in itself, Bernard and George's attempt to replace the two consultants with two others no better suited for this mission just made matters worse. Anthony finally began to catch what was happening.

- In addition, Bernard and several other consultants were exposed for their lack of knowledge and experience with the ERP software solution this company selected.

- Unfortunately, some people issues flared among financial project team members during their To Be development work sessions.

- At the same time, some project process issues flared among the consultants that Bernard could not resolve despite his repeated attempts to do so.

- Just when it seems that things could not get any worse for Anthony and this project, the company CEO announced a merger with a rival competitor that further complicated the situation for this endeavor. The uncertainty of ever completing this mission successfully was now at its pinnacle.

- Anthony had learned many hard lessons to this point, but he still had no idea how to get things righted.

In the End, chapter 15, Anthony was near the brink when an old college friend, Larry, contacted him.

- Anthony reached out to Larry for this project in a desperate attempt to change its course for the better.

- It turned out that Larry was a project management consultant and a certified Project Management Professional (PMP).

- As a result, Anthony asked his old friend to provide him with a project evaluation to include particular recommendations. Larry slowly got involved and began to contribute with his first review and next ideas for improving this important endeavor for this company.

- Then, by a stroke of good fortune, the merger with their rival was called off. Thus, one more impediment is removed as a challenge to this effort.

- Slowly and surely, as Larry's involvement increased, the project began to be reworked. The project plan and budget were modified to reflect what had already successfully been achieved to date and what remained to be done. Then, as more time passed with Larry supporting Anthony, the project got back on track and began to move forward to its culmination.

- The project took longer, resulting in higher costs, but over time, it achieved many worthwhile benefits for this company. The company CEO was pleased with the outcome, especially given the project's earlier direction. He asked Larry to return one year later to perform a brief post audit.

- Meanwhile, Anthony became a PMP and an active member of his local Project Management Institute (PMI) chapter.

- Larry's post audit revealed even better than expected results from the new ERP software's deployment, which only increased the company CEO's appreciation of it and of Anthony and Larry.

- A short time later, the CEO created a Project Management Office (PMO) for this company and appointed Anthony its manager.

This large fable has a happy conclusion, but it clearly did not start that way. Both In the Beginning and In the Middle, many classic project mistakes were shown. To that point, the project was destined for failure, if not outright collapse. However, with a PMP's help, the project altered course and began to take a new path toward success on several levels.

Anthony was fortunate when Larry first reached out to him, but he listened and then acted on this information with the support of the company CEO. He was a dedicated and hardworking person who quickly recognized the value of project management and, thus, Larry. So, too, did the company CEO. As a result, changes were set up rapidly so the endeavor could proceed without wasting any more time and money. Everyone fell in line as things became clearer and more focused. They began to be a project team instead of a collection of individuals, and

they further realized the project was both real and
serious for the future of this company.

CHAPTER SEVENTEEN

A Summary

Project management is one of the fastest growing global professions of our times. As such, more and more people worldwide are becoming project managers by choice or otherwise. This is true of my entry into project management. I had this role long before I ever knew it was a profession, with a job title, to begin with. I know that I am not alone in this regard. Like so many others, I made my share of mistakes and trusted in others who did not always deserve my trust, yet alone that of my various employers and clients. I jumped into the opportunities offered to me and tried to do my best. Sometimes, it worked out better than at other times.

Further, no one up to this day some twenty years later ever told me about the three key challenges of project management. I had to discover them for myself over time, and then with extensive personal research, I understood that one of these three was far more significant than the other two were. What I refer to is the following:

- People challenges—having to do with people interacting under stressful project conditions because of constrained schedules, budgets, and resources;

- Process challenges—having to do with various choices for applying different approaches, tools, techniques, and methods to a particular project effort.

- Technological challenges—having to do with handling new and complex technical solutions.

According to the experts who examine project results, of these three challenges, people challenges cause the greatest hurdle to project managers and their projects. Project management research spanning several decades bears this out, consistently indicating that people challenges occur four times more often than process and technological challenges combined. In addition, when such experts examine how project managers spend their time, the experts have found that they spend 70% on non-value-added activities, leaving the much smaller remainder for value added ones.

There are countless project management articles and books on the process and technological aspects and far fewer on the one challenge that stands alone as the greatest of them all—people. I believe this is because process and technological challenges are viewed as much more achievable. People challenges are far tougher to solve, yet they are always there. This book sought to share a series of five short fables and one long fable to illustrate some various

forms people challenges can take. Whether it's
the CEO, an external consultant, or a company
manager who creates a people challenge for a project
manager to deal with, it presents an aspect of project
management that too many in the profession do not
seem to recognize, yet alone know how to deal with
effectively.

For the project management profession and
the tens of thousands or more who already are or
aspire to be team leaders and project managers,
this challenge needs to be faced. It is real and
present in every project wherever or whenever
on the planet it is handled. Like the creation of
improved project management methods and tools
and technology, people challenges require attention,
study, and effective solutions. People who become
or choose to be team leaders and project managers
need to understand this. There is far more to being
a successful project manager than just mastering
MS Project, a budget, and ERP systems. Handling
a project schedule, budget, quality standards, and
scope are crucial success factors in themselves. But
these skills can be taught and learned with some
effort. People challenges too often surprise and
overwhelm good project managers, although dealing
with these typical project management factors
should be expected.

What is different about people challenges
in project management begins with the nature
of projects themselves. They are unique efforts
performed for some well-defined purpose and are
progressively developed over a certain period. Unlike
other forms of work, project management involves
assembling a team of experts and then having them

work together under stressful circumstances. The project manager has no line authority over the people who often have never worked together. When you combine all this with the three key challenges previously mentioned, you have a situation that at times seems to border on the impossible.

Being a project manager entails dealing with all this daily over the endeavor's life cycle. Above all else, people challenges make the job satisfying and terrifying. To mentor people to create a better result and to see them blossom because of their participation is exhilarating. Similarly, stress can bring out the worst in human relationships—fear, hate, lying, avoidance, shutdown, and so on. There are times when a project manager will wonder why he or she ever signed up for such punishing duty.

However, this is just the tip of the iceberg. Being a project manager is a 24/7 responsibility during the effort. You eat, sleep, and dream about the project every day and night. Peace and quiet is taken away in an instant, and it usually occurs at the least opportune moment. One minute, you are cruising along, and the next, you hit an obstacle. The job is not for perfectionists or the faint of heart. Despite it all, it is both doable and possible to achieve. It requires certain knowledge and skills, plus patience, persistence, and creativity.

Projects are often started to change one's circumstances and conditions. In today's highly competitive global economy, change tends to come frequently and cuts deep. It often entails using new and challenging technology such as ERP software, which affects significantly an entire business enterprise. The changes such endeavors bring with them are many and large in their breadth and scope. Many people are affected in many ways. Again, because people are involved, those leading the drive for such change will find the going rough and uncertain.

Just as with people challenges in project management, breaking down any resistance to change that certain people throw up takes much courage, dedication, focus, understanding, and again, creativity. There is no one way or any certain path to pursue in this journey. Again, despite all the challenges, risks, and efforts required, the job is possible to achieve and successfully.

Dealing with people is challenging. Dealing with resistance to change is difficult. When you put these two together, the task can seem insurmountable. Just as a journey of a thousand miles begins by taking the first step and then the next, so, too, do the efforts you need to expend each day of the project's life cycle.

Some days, you feel as if you are making no progress or even worse, going backward. Then, there are days in which you can do no wrong. But if you clearly know where you want to get to and stick to it, you are more likely to get there. Project management is a disciplined profession that provides project managers with skills and tools to apply to help them in this worthy effort. No one can guarantee success; however, by taking advantage of others' experience, you can enhance the odds. That's what it's all about and why this book relied on stories.

Last, despite the hardships you must take at times along the way, it is all worth it in the end. Those who endure it all and successfully reach the end of the road know true joy. It is not easy, nor quick, nor painless. You could well become exhausted. If it feels as if you are running a marathon, you are. But through training and development, having a clear road map to study and follow, and by being flexible within certain boundaries, you can achieve success. Many information technology projects historically have had an awful record of accomplishment; however, things are slowly improving. This is not just due to better methods, tools, and techniques. It is also because the project management profession finally realizes that people challenges are real. Recognition of a problem is the first crucial step in developing solutions.

CHAPTER EIGHTEEN

Richard C. Bernheim's Project Management Quotations

- *"A project manager obtains disproportionate credit with disproportionate blame."*
- *"A project manager lives a project 24/7, from its start to its completion."*
- *"Project management is first all about leading people and then secondarily about applying process (methods and tools) and technology. Leading people is challenging as well as rewarding."*
- *"Projects need to be managed, but to be successful, they require true leadership."*
- *"A project-driven change, such as building a bridge or tunnel where none previously was, is more likely to be widely accepted, whereas changing a person's job by altering his or her work methods and tools is not. I believe this*

169

is due to the truth that the bridge or tunnel is viewed as some kind of betterment, whereas a change to a person's job is initially seen as a threat."

- "There are two key dimensions in the real-world practice of project management:

  o The challenges of working with people in a project environment.

  o Projects most often are used to drive some sort of change, and people tend to resist change because of fear of the unknown."

- "Change is difficult to achieve because the people's resistance to it is like trying to defy gravity. It takes much intense thrust to be successful."

Project versus Change Management

Project Management:

- Task focused
- Process—follows the five project phases
- Tools used—statement of work, project charter, business case, work breakdown structure, budget, schedule, and resource allocation

Change Management:

- People oriented
- Process—concentrates on both the organization and the person
- Tools used—readiness assessment, communication plans, and training plans

What They Have in Common—the goal of moving us from an existing condition (As Is) to a much better one (To Be).

About the Author

Richard Bernheim has been a senior management consultant in three of the top four consulting firms (Ernst & Young, Deloitte & Touche, and Price Waterhouse) where he has specialized in implementing financial and cost management information systems such as SAP R/3. He was featured in the 1987 (25th) edition of *Who's Who in the Finance Industry in America.*

A popular educator and speaker, he has been a part-time professor of financial management and project management at Montgomery County Community College, Temple University, and University of Phoenix. He has also been a featured speaker at various conferences, at PMI chapter meetings, and at other professional organizations since 1983.

An accomplished writer, Bernheim has authored 10 published manuscripts for professional organizations and has proofread and revised several college-level financial and cost accounting textbooks.

He holds a BA in political science, an MBA in finance, a PMP from the Project Management Institute, and is certified in various SAP R/3 implementation methodologies and ASAP Project Management.

To obtain additional information, including related materials, please check out the author's Website at the following Internet address:

www.BeaSmartPM.com

Did you like this book?

If you enjoyed this book, you will find more interesting books at

www.MMPubs.com

Please take the time to let us know how you liked this book. Even short reviews of 2-3 sentences can be helpful and may be used in our marketing materials. If you take the time to post a review for this book on Amazon.com, let us know when the review is posted and you will receive a free audiobook or ebook from our catalog. Simply email the link to the review once it is live on Amazon.com, with your name, and your mailing address—send the email to orders@mmpubs. com with the subject line "Book Review Posted on Amazon."

If you have questions about this book, our customer loyalty program, or our review rewards program, please contact us at info@mmpubs.com.

People-Centric Project Management

Research has shown that most projects fail not because of technological issues but, rather, issues related to people such as a lack of (or poor) communication, inadequate leadership, unclear lines of authority, and poor motivation. Dealing with individual behaviors, organizational culture, and internal politics can distract a project manager from value-added activities, leading to overwork, stress, and overall project chaos.

Using three case studies to frame the analysis, this book illustrates the good, the bad, and the ugly sides of how people can affect a project's outcomes. It also explains the root causes of people issues and how to best deal with them while managing a project.

Every project manager faces challenges when dealing with people. Read this book and learn how better to prevent these issues and how to resolve them when they do arrive.

Available in paperback format. Order from your local bookseller or directly from the publisher at

http://www.mmpubs.com/

Top-Gun Project Managers: Reaching the Top of the PM Profession

Ever wonder why some people's careers rocket upwards while yours feels stuck, with you repeating the same drudgery day after day? If you are a project manager (or are interested in becoming one) then this book was written just for you. It shares eight strategies that you can use when plotting your career path—the trajectory of your own rocket—to help you reach the stratospheric levels of the profession.

The author of this book, Richard Morreale, is one of the top project managers in the world, specializing in turning around some of the nastiest, largest, troubled projects you will ever find. Morreale's career has spanned a wide range of projects from working as part of the Apollo Program Team, helping to put men on the moon, to working as part of the team that computerized the UK Income Tax System.

Read this book to learn how you too can copy Morreale's career success—with these eight strategies, the sky's the limit!

Available in paperback format. Order from your local bookseller or directly from the publisher at

http://www.mmpubs.com/

Managing Smaller Projects: A Practical Approach

So called "small projects" can have potentially alarming consequences if they go wrong, but their control is often left to chance. The solution is to adapt tried and tested project management techniques.

This book provides a low overhead, highly practical way of looking after small projects. It covers all the essential skills: from project start-up, to managing risk, quality and change, through to controlling the project with a simple control system. It cuts through the jargon of project management and provides a framework that is as useful to those lacking formal training, as it is to those who are skilled project managers and want to control smaller projects without the burden of bureaucracy.

Read this best-selling book from the U.K., now making its North American debut. *IEE Engineering Management* praises the book, noting that "Simply put, this book is about helping people control smaller projects in a logical and effective way, and making the process run smoothly, and is indeed a success in achieving that goal."

Available in paperback format. Order from your local bookseller or directly from the publisher at

http://www.mmpubs.com/msp

The History of Project Management

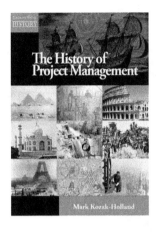

The Pyramid of Giza, the Colosseum, and the Transcontinental Railroad are all great historical projects from the past four millennia. When we look back, we tend to look at these as great architectural or engineering works. Project management tends to be overlooked, and yet its core principles were used extensively in these projects.

This book takes a hard look at the history of project management and how it evolved over the past 4,500 years. It shows that "modern" project management practices did not just appear in the past 100 years but have been used — often with a lot of sophistication — for thousands of years.

Examining archaeological evidence, artwork, and surviving manuscripts, this book provides evidence of how each of the nine knowledge areas of project management (as shown in PMI's PMBoK® Guide) have been practiced throughout the ages.

As readers explore the many case studies in this book, they will discover fascinating details of innovative projects that produced many of our most famous landmarks and voyages of discovery.

Available in paperback format. Order from your local bookseller or directly from the publisher at

http://www.mmpubs.com/